Springer Theses

Recognizing Outstanding Ph.D. Research

Aims and Scope

The series "Springer Theses" brings together a selection of the very best Ph.D. theses from around the world and across the physical sciences. Nominated and endorsed by two recognized specialists, each published volume has been selected for its scientific excellence and the high impact of its contents for the pertinent field of research. For greater accessibility to non-specialists, the published versions include an extended introduction, as well as a foreword by the student's supervisor explaining the special relevance of the work for the field. As a whole, the series will provide a valuable resource both for newcomers to the research fields described, and for other scientists seeking detailed background information on special questions. Finally, it provides an accredited documentation of the valuable contributions made by today's younger generation of scientists.

Theses are accepted into the series by invited nomination only and must fulfill all of the following criteria

- They must be written in good English.
- The topic should fall within the confines of Chemistry, Physics, Earth Sciences, Engineering and related interdisciplinary fields such as Materials, Nanoscience, Chemical Engineering, Complex Systems and Biophysics.
- The work reported in the thesis must represent a significant scientific advance.
- If the thesis includes previously published material, permission to reproduce this must be gained from the respective copyright holder.
- They must have been examined and passed during the 12 months prior to nomination.
- Each thesis should include a foreword by the supervisor outlining the significance of its content.
- The theses should have a clearly defined structure including an introduction accessible to scientists not expert in that particular field.

More information about this series at http://www.springer.com/series/8790

Li Li

Modeling the Fate
of Chemicals in Products

Doctoral Thesis accepted by
Peking University, Beijing, China

 Springer

Author
Assist. Prof. Li Li
School of Community Health Sciences
University of Nevada Reno
Reno, NV, USA

Supervisor
Prof. Jianxin Hu
Department of Environmental Sciences
Peking University
Beijing, China

ISSN 2190-5053   ISSN 2190-5061   (electronic)
Springer Theses
ISBN 978-981-15-0578-2   ISBN 978-981-15-0579-9   (eBook)
https://doi.org/10.1007/978-981-15-0579-9

This Springer imprint is published by the registered company Springer Nature Singapore Pte Ltd.
The registered company address is: 152 Beach Road, #21-01/04 Gateway East, Singapore 189721, Singapore

Supervisor's Foreword

A substantial use of chemicals is a double-edged sword. Chemicals are magical because they have shaped our modern lives and saved countless lives; on the other hand, however, enormous chemicals can exert negative impacts on the environment and human health. Dating back to the 1960s, the pioneering work *Silent Spring* by Rachel Carson for the first time revealed how organochlorine pesticides such as dichlorodiphenyltrichloroethane (DDT) influence the ecosystem. In 1992, *Agenda 21* encouraged "environmentally sound management of toxic chemicals" by 2000 and proposed "expanding and accelerating international assessment of chemical risks" as an important program area. This commitment was then restated and reemphasized in the World Summit on Sustainable Development in 2002, which called for "minimization of significant adverse effects on human health and the environment, using transparent science-based risk assessment procedures and science-based risk management procedures" by 2020 (United Nations, *Report of the World Summit on Sustainable Development*, Johannesburg, 2002). The importance of environmentally sound management of chemicals can never be overemphasized given that three out of the 17 Sustainable Development Goals set by the *2030 Agenda for Sustainable Development* are related to the chemical issue (United Nations, *Transforming our world: the 2030 Agenda for Sustainable Development*, New York, 2015).

It is the properties of chemicals and associated products that determine the transport, transformation and evolution of chemicals in the human socioeconomic system and the environment. This is a fundamental rule. However, it is rather challenging to predict the fate of chemicals in products in such a complex nexus because the physicochemical properties and lifecycle features vary remarkably between chemicals and between products. For this reason, a pressing issue in environmental sciences is to develop methodological approaches that seek to characterize the accumulation, transformation and emissions of chemicals in the "total" environment. This thesis makes an exceptional contribution to realizing this goal, by developing an integrative, dynamic and mechanistic anthropospheric and environmental fate modeling framework, which makes it possible to link the fate of chemicals in the "total" environment with the physicochemical properties of chemicals and the lifecycle features of products. In this regard, this thesis no doubt provides a fascinating example for this effort. I do hope the results and conclusions presented in this thesis can benefit research and practices in chemical management, cleaner production, environmental systems analysis, and so on.

I have been enjoying the pleasant memory with Dr. Li Li ever since he joined my research group. During our frequent exchange of ideas on the issue of chemicals in products, I witnessed his profound academic background in environmental sciences and insightful understanding of the issues of chemicals in products. Of course, like all other theses, this thesis is a balance, or a compromise, between his pursuit of perfection in science and the best scientific understandings available to date. I believe that students and scholars in the fields of environmental sciences and chemical management, or readers in other fields, will benefit from this work—At least I know that I have benefited a great deal from his work.

Beijing, China   Prof. Jianxin Hu
March 2018

Abstract

As an integrated part of an industrialized economy, innumerable chemicals in products (CiPs) offer modern comfort and convenience but also pose unintended risks for the environment, wildlife as well as human health. Unlike industrial chemicals or unintentionally generated byproduct contaminants, which often enter and travel through the abiotic physical environment immediately after generation, CiPs can undergo transport and transformation in the human socioeconomic system (the anthroposphere) throughout product lifespan and waste disposal phases, as well as accumulation in consumer products in service (i.e., in-use stock) and dumps and landfills (i.e., waste stock). Therefore, modeling the behavior and fate of CiPs in both the anthroposphere and the physical environment is essential if we want to understand and assess their environmental risks so as to achieve better environmentally sound management of CiPs and waste.

In this book, a dynamic, mechanistic anthropospheric fate model, CiP-CAFE (*Chemicals in Products—Comprehensive Anthropospheric Fate Estimation*), was first developed for quantifying the flows and stocks of CiPs in the anthroposphere. The model was then coupled with a dynamic, mechanistic environmental fate model, BETR-Global, to form a coherent and compatible modeling framework for establishing a continuum from production to concentrations in environmental compartments. As case studies, the modeling framework was applied to simulate (i) the dominant route by which polychlorinated biphenyls disperse across the global environment, (ii) degradation of fluorotelomer-based polymers in waste stocks and its contribution to the global occurrence of fluorotelomer alcohols and perfluoroalkyl carboxylates, (iii) variability in the stereoisomer profile of hexabromocyclododecane as a result of isomerization in the anthroposphere and physical environment, and (iv) influence of waste disposal options on the long-term hexabromocyclododecane emissions and contamination in China. General conclusions summarized from the cases apply to a wider range of CiPs.

The modeling results demonstrate that the longer a CiP remains in in-use stock (i.e., product lifespan) and waste stock (i.e., degradation half-life), the more it is enriched in the two stocks, and the later the two stocks peak. CiPs keep moving from in-use stock to waste stock throughout the product service life. However, both in-use and waste stocks are transient storages of CiPs in the anthroposphere. In the long term, all CiPs in the anthroposphere will eventually be either liberated into the physical environment or decomposed in the anthroposphere. The results indicate that the extent of the global dispersal of CiPs caused by humans (i.e., via international trade of products and wastes) is more remarkable than that occurring in the physical environment (i.e., via atmospheric and oceanic advection). In addition, ongoing interspecies transformation renders composition of a CiP observed in the physical environment substantially distinct from that measured in its technical mixtures, which might be linked to elevated environmental and health risks. The results also illustrate that regulative actions implemented on a substance level are more effective in minimizing CiP emissions than those on a material or waste level, which warrants an active, precautionary strategy in the environmentally sound management of CiPs and waste. Crucially, the modeling used here reveals that accumulation of CiPs in two anthropospheric stocks acts as a "time buffer" that extends and possibly postpones the CiP emissions to a long period of decades or even centuries.

In conclusion, this book establishes a methodological and theoretical basis for a better understanding of the fate and risk of CiPs in the total environment. It also provides scientific guidance to both industry and policymakers with respect to green-chemistry design, cleaner production and environmentally sound management of chemicals.

Keywords Chemicals in products · Chemical fate · Substance flow analysis · Fugacity model · Anthroposphere

Parts of this thesis have been published in the following journal articles

Li, L., Weber, R., Liu, J., Hu, J., Long-term emissions of hexabromocyclododecane as a chemical of concern in products in China, *Environ. Int.* 2016, 91: 291–300 (Reproduced with Permission. Copyright (2016) Elsevier).

Li, L., Wania, F., Tracking chemicals in products around the world: introduction of a dynamic substance flow analysis model and application to PCBs, *Environ. Int.* 2016, 94: 674–686 (Reproduced with Permission. Copyright (2016) Elsevier).

Li, L., Liu, J., Hu, J., Wania, F., Degradation of fluorotelomer-based polymers contributes to the global occurrence of fluorotelomer alcohol and perfluoroalkyl carboxylates: A combined dynamic substance flow and environmental fate modeling analysis, *Environ. Sci. Technol.* 2017, 51: 4461–4470 (Reproduced with Permission. Copyright (2017) American Chemical Society).

Li, L., Wania, F., Elucidating the variability in the hexabromocyclododecane diastereomer profile in the global environment, *Environ. Sci. Technol.* 2018, 52: 10532–10542 (Reproduced with Permission. Copyright (2018) American Chemical Society).

Acknowledgements

I am deeply indebted to and would like to express my sincere gratitude to

Prof. Jianxin Hu for offering me an independent, free, diverse and comfortable research atmosphere, as well as enlightened attitudes towards international collaborations.

Prof. Frank Wania (University of Toronto Scarborough, Canada) for his thoughtful, inspiring and heuristic mentorship and discussion, infectious enthusiasm for sciences, and unfailing generous support for my academic career.

Dr. Jianguo Liu (Peking University, China) for introducing me to the promising and fruitful research area of chemicals in products, and unselfishly providing me tremendous opportunities for international collaborations.

Dr. Jianhua Xu (Peking University) for her professional training of me in research methodologies, academic integrity and academic writing, which equips me with powerful tools to survive in academia and to appreciate the beauty of sciences.

Dr. Hidetoshi Kuramochi (National Institute of Environmental Sciences, Japan) and Dr. Knut Breivik (Norwegian Institute for Air Research, NILU) for their insightful discussion on the physicochemical properties and environmental fate of chemicals; Dr. Roland Weber (POPs Environmental Consulting, Germany) for his insightful discussion on the industrial ecological features and anthropospheric fate of products.

Dr. Golnoush Abbasi (NILU) for her detailed feedback and suggestions during the use and refinement of the CiP-CAFE model.

Dr. Xinghua Qiu and Dr. Weiguang Xu (Peking University), and Dr. Jun Huang (Tsinghua University, China), for their helpful comments on the early version of the doctoral thesis.

Dr. Alessandro Sangion and Sivani Baskaran (University of Toronto Scarborough) for their helpful comments and careful editing during preparation of this book.

Mr. Shengfang Mei (China Association of Fluorine and Silicone Industry), Dr. Shuzhong Wang (Asahi Glass), Dr. Jason Zheng (Chemours), and Dr. Ichikawa Tosei (Daikin Fluorochemicals), for providing information on production and use of perfluoroalkyl substances.

Mr. Zhengmao Zhou and Dr. Lijun Qian (Beijing Institute of Technology, China), and Mr. Qingjun Meng (China Plastics Processing Industry Association), for providing information on production and use of hexabromocyclododecane in China.

Financial support by the National Natural Science Foundation of China (grant no. 21577002) for the work described in Chap. 4.

Scholarships provided by the China Scholarship Council and Shanghai Tongji Gao Tingyao Environmental Science and Technology Development Foundation for supporting my collaboration with Dr. Frank Wania.

Talented and intelligent friends who made the journey toward a Ph.D. degree memorable: Pengju Bie, Hong Cai, Chengkang Chen, Dr. Yian Dong, Jiarui Han, Zhi Huang, Dr. Shenglan Jia, Dr. Torsten Juelich, Huan Li, Huanhong Li, Yan Lin, Ziwei Mo, Dr. Shenshen Su, Bowen Ti, Dr. Qiang Wang, Ruoyu Wang, Ziyuan Wang, Tian Xia, Dr. Qiaoyun Yang, Mengqi Yu, Mengke Zhao, Ximeng Zhao and Yan Zhao from Peking University, and Dr. James Armitage, Tife Awonaike, Dr. Matthew Binnington, Yuhao Chen, Ying Duan Lei, David McLagan, Joesph Okeme, Abha Parajulee, Dr. Li Shen, Qianwen Shi, Yuchao Wan, Dr. Chen Wang and Stephen Wood from the University of Toronto, as well as many others.

I am especially grateful to my parents, Jinjun Li and Guanrong Shi, and my wife, Yuchao Lu, for their unconditional and unwavering love, understanding and support.

Contents

Part I Methodologies

**1 Introduction: Modeling the Fate of Chemicals in Products
in the Total Environment**   3
  1.1 Chemicals in Products   3
  1.2 The Total Environment   5
  1.3 Modeling the Fate of CiPs in the Total Environment   7
      1.3.1 Modeling the Fate of Products in the Anthroposphere   8
      1.3.2 Modeling the Fate of Chemicals in the Environment   13
  1.4 Research Needs and Objectives   17
      1.4.1 Need for an Integrative Modeling Framework   17
      1.4.2 Need for a Temporally Resolved Modeling Framework . . .   18
      1.4.3 Need for a Mechanistic Modeling Framework   19
      1.4.4 Research Objectives   21
  1.5 The Structure of the Book   21
  References   21

**2 Developing Models for Tracking the Fate of Chemicals
in Products in the Total Environment**   27
  2.1 Analogy Between Chemical Anthropospheric
       and Environmental Fate Modeling   27
  2.2 CiP-CAFE: Anthropospheric Fate Modeling   29
      2.2.1 Structure and Conventions   29
      2.2.2 Calculation Modes   32
      2.2.3 Configuration and Parameterization   34
      2.2.4 Model Outputs   39
  2.3 BETR-Global Model: Environmental Fate Modeling   39
      2.3.1 Structure and Conventions   39
      2.3.2 Emissions: A Bridge Between CiP-CAFE
             and BETR-Global   41

2.4   Summary   41
References   41

Part II   Case Studies

**3   Global Long-Term Fate and Dispersal of Polychlorinated
      Biphenyls**   47
      3.1   Introduction   47
      3.2   Methods and Data   48
             3.2.1   Properties of PCB Congeners   48
             3.2.2   Production, Applications and Interregional Trade   49
             3.2.3   Emission, Waste and Decomposition Factors   50
             3.2.4   Disposal of PCB-Containing Solid Waste   50
      3.3   Evaluation of Modeling Performance   51
      3.4   Overview of the Fate of PCBs in the Anthroposphere   52
      3.5   Dispersal of PCBs in the Global Anthroposphere and
             Environment   56
      3.6   Summary   58
      References   58

**4   The Degradation of Fluorotelomer-Based Polymers Contributes
      to the Global Occurrence of Fluorotelomer Alcohols
      and Perfluoroalkyl Carboxylates**   63
      4.1   Introduction   63
      4.2   Method and Data   64
             4.2.1   Applications, Chemicals and Terminology   64
             4.2.2   Modeling Strategy   66
      4.3   Temporal Evolution of Stocks and Releases   69
      4.4   Rising Contributions of FTP Degradation to Global FTOH
             Releases   72
      4.5   FTP Degradation in Waste Stocks as a Long-Term Source
             of PFCAs   74
      4.6   Summary   75
      References   76

**5   Elucidating the Variability in the Hexabromocyclododecane
      Diastereomer Profile in the Global Environment**   79
      5.1   Introduction   79
      5.2   Method and Data   81
             5.2.1   Properties of HBCDD Diastereomers   81
             5.2.2   Production, Applications and Interregional Trade   81
             5.2.3   Emission and Waste Factors   83
             5.2.4   Disposal of HBCDD-Containing Solid Waste   84

	5.2.5	Quantification of Isomerization	84
	5.2.6	Uncertainty Scenarios	87
5.3		Evaluation of Modeling Performance	88
5.4		Global Emissions of HBCDD Diastereomers	89
5.5		Occurrence of HBCDD Diastereomers in the Global Environment	92
5.6		Summary	94
References			95

6 Effective Management of Demolition Waste Containing Hexabromocyclododecane in China 99

6.1		Introduction	99
6.2		Methods and Data	100
	6.2.1	Modifications for the China-Specific Situation	100
	6.2.2	Disposal Scenarios of HBCDD-Containing Demolition Waste	101
	6.2.3	Evaluation of Emission Mitigation Performance	104
6.3		Overview of Stocks and Emissions in Mainland China	105
6.4		Emission Mitigations in Different End-of-Life Disposal Scenarios	107
6.5		Summary	109
References			110

Part I
Methodologies

Chapter 1
Introduction: Modeling the Fate of Chemicals in Products in the Total Environment

Abstract Academic and public interest has been poured into "chemicals in products (CiPs)", that is, the chemicals of environmental and health concern intentionally added in industrial or consumer goods for additional functions. This chapter gives a comprehensive overview of the fate of CiPs in the "total" environment, that is, a system comprising the anthroposphere (human socioeconomic activities) and the physical environment (the totality of abiotic setting in which organisms and humans live) as two interconnected and interdependent components. Currently, industrial ecologists have well elucidated the anthropospheric fate of products whereas environmental chemists have well elucidated the environmental fate of chemicals. Nevertheless, there is still a need for an integrative, temporally resolved, and mechanistically sound modeling framework for chemicals plus products as an organic whole in the CiP issue.

1.1 Chemicals in Products

The past decades have witnessed the intensified use of >100,000 synthetic chemicals in products/articles (CiPs) in all aspects of our lives, e.g., in homes, workspaces and schools, for modern comfort and convenience [1]. For example, hundreds of flame-retardants are embedded in a diverse range of industrial and consumer products, such as electronics, vehicles, furniture, building materials and textiles, to decelerate or prevent the development of ignition [2, 3]. Such "chemical intensification" of our economy has two dimensions [4]: First, an increasing number of synthetic, or human-made, chemicals are replacing natural materials in industrial and consumer products; in the meanwhile, the ever-growing chemical industry endows the synthetic chemicals with a myriad of novel functions and thus creates new consumer needs.

While the use of the majority of the >100,000 synthetic chemicals is considered to be "safe", a considerable fraction of them has been realized to be "troublesome" for the ecosystem and human health. A notorious example is endocrine-disrupting chemicals that alter the human endocrine functions and potentially cause cancers, obesity, diabetes and other diseases. Names on this list to date include polychlorinated

© Springer Nature Singapore Pte Ltd. 2020
L. Li, *Modeling the Fate of Chemicals in Products*, Springer Theses,
https://doi.org/10.1007/978-981-15-0579-9_1

biphenyls (PCBs), polybrominated diphenyl ethers (PBDEs), hexabromocyclododecane (HBCDD), perfluoroalkyl carboxylates (PFCAs) and sulfonates (PFSAs), and short-chain chlorinated paraffins (SCCPs), and are still expanding. It is estimated that the exposure of the general populations to these endocrine-disrupting chemicals have led to annual disease costs of 340 and 209 billion US dollars in the United States [5] and European Union [6], respectively. As such, the chemical intensification of the economy is double-edged: we may suffer ever-increasing adverse health and environmental risks while we are enjoying the huge social and economic benefits arising from the daily routine use of the myriad of chemicals [1]. Environmentally sound management of these synthetic chemicals, therefore, creates a new social need and becomes a consensus between industries, academia, and regulatory agencies. In 2006, an international platform named the Strategic Approach to International Chemicals Management (SAICM) was established with the goals to achieve "the sound management of chemicals throughout their life cycle" and to promote that "chemicals are used and produced in ways that lead to the minimization of significant adverse effects on human health and the environment" by 2020 [7]. In 2015, the 2030 Agenda for Sustainable Development adopted by the UN General Assembly restates the need for efforts to achieve "the environmentally sound management of chemicals and all wastes throughout their life cycle" and to reduce their "release to air, water and soil in order to minimize their adverse impacts on human health and the environment" by 2020. Accordingly, the ultimate aim of this book is to help industries, academia, and regulatory agencies to achieve these goals toward reducing environmental and health risk of synthetic chemicals and creating a chemical-safe world.

Despite the relatively long recognition of synthetic chemicals as an environmental issue, it is not until very recently that chemicals and products/articles were studied as a whole. The combination of chemicals and products/articles complicates the issue of synthetic chemicals because it additionally introduces the lifecycle of products and the behavior of products in human society into the chemical contamination issue. The term "CiP" was originally coined by the Overarching Policy Strategy of the SAICM, which called for that "information on chemicals throughout their life cycle, including, where appropriate, *chemicals in products*, is available, accessible, user friendly, adequate and appropriate to the needs of all stakeholders" [7]. However, the Overarching Policy Strategy did not provide the definitions and scopes of "chemicals" and "products". In order to better address the CiP issue, the International Conference on Chemicals Management (ICCM) nominated CiPs as a global "emerging policy issue" in its second session in 2009. The third session of the ICCM launched a multistakeholder collaborative project named "Chemicals in Products Programme", with the aim of "facilitating and guiding the provision and availability of, and access to, relevant information on chemicals in products among all stakeholder groups" (ICCM Resolution III/2). The outcome of the project then became a voluntary framework for all SAICM stakeholders in the fourth session of the ICCM in 2015 [8]. Several national regulations, such as the Consumer Product Safety Improvement Act (CPSIA) in the United States and the Restriction of Hazardous Substances (RoHS) in the European Union, are also tackling the CiP issue.

According to the SAICM CiP framework, a "product" is defined as "an object that during production is given a special shape, surface or design which determines its function to a greater degree than its chemical composition" [8], which is essentially the definition of an "article" in other regulatory documents such as the Registration, Evaluation, Authorisation and Restriction of Chemicals (REACH) [9]. Examples of products include diapers, cars, insulation boards and televisions. For detailed definitions and examples of products (or articles), readers are recommended to consult the EU Guidance on Requirements for Substances in Articles [10]. The SAICM CiP framework also treats "products" as an alternative to "goods", i.e., tangible physical entities such as "textiles, furniture, construction materials, electronics, household items and other consumer goods" [8]. Note that, to be distinguished from "articles" defined in the REACH, a fraction of literature refers to personal care and cleaning products (e.g., shampoo, detergents and toothpaste) as "products", for which functions are largely dependent on their chemical compositions. In this book we keep consistent with the SAICM terminology.

An important issue is at which level a "product" should be defined: the product model (products from a specific company or brand), product form (a series of similar product models), or product class (the category of products) [11]. For example, videocassette recorders (product class) can be a two-, four- or six-head (product form), and each of them can further come from different manufacturers (product model) [11]. When talking about the use of CiPs in the entire society, we often regard the product as an equivalent to the product class and do not differentiate between specific product forms and models.

In the meantime, according to the SAICM CiP framework, the term "chemical" specifically refers to the chemical substances of environmental and/or health concern [8]. This definition encompasses chemical groups such as (i) persistent, bioaccumulative and toxic (PBT) substances, or very persistent and very bioaccumulative (vPvB) substances, and (ii) substances that are carcinogenic, mutagenic, and toxic for reproduction (CMR). In some cases [10], attention should also be paid to chemical substances present in products with a mass fraction exceeding 0.1%, irrespective of whether they are found to be hazardous. While inorganic chemicals, such as heavy metals, can also be hazardous, we limit our scope to organic chemicals. In this book, we do not discuss chemical components unintentionally generated and/or present in products, such as dioxin in contaminated toys.

1.2 The Total Environment

Before entering various environmental compartments such as air, water and soil, a chemical can reside in the human socioeconomic system for a long time because of human activities such as chemical synthesis, manufacturing of chemical-containing products, product use and waste handling. Organisms and humans are exposed to a chemical not only from those environmental compartments but also during these human activities. Therefore, we need to extend the definition of an "environment"

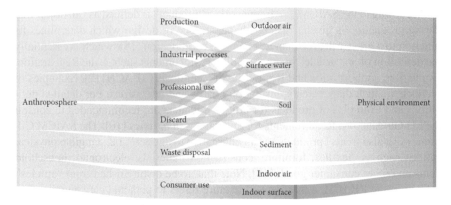

Fig. 1.1 Components of the total environment (the anthroposphere and physical environment) and compartments

from the *"physical environment"*, i.e., all abiotic physical setting in which we live, to a *"total"* environment, by incorporating an *"anthroposphere"* that includes human socioeconomic activities occurring throughout the lifetime of a chemical (Fig. 1.1). The anthroposphere is also termed as the "technosphere", or "physical economy" (as opposed to the "financial economy") [12]. In the literature, the anthroposphere denotes the part of the total environment "made or modified by humans and used for their activities [13]", which is an "aggregate of all individual human lives, actions and products [14]". By these definitions, we can roughly treat the anthroposphere as a nexus of all human technological activities embedded in the supply chain (from synthesis to end consumers), use phase (from end consumers to waste) and waste disposal. In addition to the anthroposphere and physical environment, one can also include a "biosphere" in the total environment, which includes all living organisms and humans connected by the food webs. We do not discuss the biosphere in this book.

The anthroposphere and physical environment can be further divided into multiple interconnected "compartments" (Fig. 1.1). The anthroposphere comprises (i) lifecycle stages of products, such as production, formulation, professional use, consumer use and in service, and (ii) waste disposal activities, such as engineered landfill, dump and incineration of waste. The physical environment comprises various environmental media, such as surface water, indoor and outdoor air, soil, and sediment, through which living organisms and humans can take chemicals into their bodies. Strictly speaking, the physical environment also contains the lithosphere, i.e., a part of the environment that has no interactions with the other environmental compartments or living organisms [12].

These compartments can be viewed as containers or reservoirs to store chemicals for a certain period. The mass of a chemical is a good indicator to gauge such storage. For instance, flame-retardant molecules are brought into the anthroposphere once they are synthesized. If added into an electronic device, the molecules can reside in the anthroposphere for years when the electronic device remains in service in our homes.

In this case, we use an "in-use stock" to describe the accumulation of chemicals in the totality of all in-service products within a region. Likewise, these flame-retardant molecules enter an engineered landfill site if an electronic device is discarded and landfilled. The molecules can reside in the anthroposphere for another several years before microorganisms degrade them in the landfill sites. We use a "waste stock" to describe that in the totality of all wastes present within a region. The storage of chemicals also occurs in environmental compartments, which is often termed as an "environmental budget" in the literature [15]. For instance, approximately 400 tonnes of PCBs were estimated to reside in the British environment at the end of the 1980s, accounting for 1% of the historical PCB sales in the UK; 93.5% of these PCBs were present in soils, followed by seawater (3.5%) and marine sediment (2.1%) [16].

A chemical can be transported between two anthropospheric or environmental components, or between components in different geographic regions within the anthroposphere or physical environment. An example for anthropospheric transport is the flow of PBDEs from the in-use stocks to waste stocks along with PBDE-containing electronics and vehicles entering the waste phase at the end of life, which was estimated to amount to 10 kilotonnes per year in North America between 2005 and 2008 [17]. A most well-known example for the environmental transport is the long-range transport of persistent organic pollutants, such as the global transport of PCBs mediated by atmospheric movement [18] and the global transport of PFCAs mediated by ocean currents [19]. The rate of such a relocation (i.e., mass per time) is termed "flow". Note that a flow can also be called as a flux in some cases, although some scientists argue that we should additionally take into account the specific area of a surface that chemicals move through when using the term "flux" [20]. Among flows a unique one is "emission", i.e., unidirectional migration of a chemical from the anthroposphere to physical environment, because it crosses the interface between anthroposphere and physical environment and links these two subsystems of the total environment.

Likewise, a chemical can be converted from its original form into another, which leads to the dissipation of this chemical. For instance, when being subject to thermal processing in the anthroposphere or exposed to light in the physical environment, HBCDD can transform from one stereoisomer to another, resulting in a stereoisomer profile completely different from that in the technical HBCDD mixture [21].

In this book, the outcome, or the destiny, resulting from the accumulation, transport and dissipation processes of CiPs in the anthroposphere and physical environment are collectively referred to as the *fate* of CiPs in the total environment, which is indicative of how a CiP is most likely to behave after its manufacturing and use.

1.3 Modeling the Fate of CiPs in the Total Environment

The environmental and health problems of CiPs are often large in spatial scale and long in temporal course, which limits the appropriateness of investigations based on fieldwork or wet-lab experiments. Therefore, the use of models become imperative and powerful. At present, investigations into the fate of *products/articles* in the *anthroposphere* are the key theme of the discipline of *industrial ecology*, whereas investigations into the fate of *chemicals* in the *physical environment* constitute the research theme of the discipline of *environmental chemistry*. The following section will present a comprehensive overview of basic notions, terminology, knowledge and information used in these two areas, which serves as the motivation of our work described in this book, as well as a key to facilitating understanding it.

1.3.1 Modeling the Fate of Products in the Anthroposphere

Industrial ecologists use material flow analysis (also called "socioeconomic metabolism") to depict the fate of products in the anthroposphere. Most of material flow analyses focus on how to achieve efficient and effective utilization of materials, how to manage and conserve beneficial resources, and how to avoid the generation and discharge of unwanted waste [22, 23]. In general, the umbrella term "material" refers to (i) entities or objects with a positive or even negative economic value, including products, merchandise, commodities, and their synonyms [22], and (ii) bulk materials, which constitute the entities and objects, such as plastics, concrete and fibers [12]. Archetypal materials include consumer durables (such as furniture and electrical appliances) [24], metals (such as iron, copper, or aluminum) [25], elements (such as chlorine) [26], and engineered nanomaterials [27]. Materials can be sold, bought, used and disposed of. Some industrial ecologists also use the term "substance" as an alternative to "material" when describing studies on (i) chemical elements such as metals and nutrient elements, and (ii) chemical compounds composed of uniform units or moieties [12, 22]. For example, in the field of global climate change, a carbon flow analysis often involves not only carbon dioxide but also other compounds containing carbon such as carbonates. However, the use of "substance" in these studies differs subtly from the characteristics of CiPs that are discussed in this book, because the "substances" are designated to be conservative, i.e., neither destroyed nor transformed in the total environment [28], whereas CiPs discussed here are organic compounds that can be subject to chemical or biological transformation in the total environment. For instance, the total amount of carbon will not be lost in a carbon flow analysis, because carbon, which is the core of a carbon flow analysis, conserves regardless of its specific speciation. In this book, we will expand the meaning of "substance" to incorporate CiPs, which can be subject to various chemical or biological transformation in the anthroposphere.

Terminologies and methodologies in material flow analysis mostly stem from the pioneering work by Robert Ayres and his collaborators from the later 1960s. The initial incentive of looking at material flows is to "view environmental pollution and its control as a materials balance problem for the entire economy" [29]. In the eyes of Robert Ayres and collaborators, the human socioeconomic system is analogous to a biological or ecological system, where different *processes*, including the accumulation, transport, transformation and dissipation, of materials are taking place all the time. Raw materials enter the human socioeconomic system, a tiny part of which is intentionally converted to and stored in *final products* and the rest becomes *residuals* due to technical limitations. Here, "residual" is an umbrella term including both obsolete materials or by-products sent to landfills, wastewater treatment plants and other waste treatment facilities ("waste"), and materials discharged into soils, rivers and other receiving environment media ("emission"). Flowing into the residual stream is inevitable because materials are inherently dissipative and tend to be readily degraded, dispersed, and emitted in the course of normal uses. That is, the efficiency of converting materials between lifecycle stages, i.e., the conversion efficiency, is not 100% in most cases; the longer the lifecycle process chain is, the lower overall conversion efficiency is likely to be. The conversion efficiency of each lifecycle stage reflects technical advancement. Because of the storage of materials, final products can be viewed as reservoirs, or *stocks*, of materials. The use, or consumption, of the final products, produces *service*, which describes the satisfaction of various human needs. The longevity of a product remaining in service is termed *product lifespan* or *product lifetime*. All final products will ultimately enter the residual stream. In all stages throughout the lifecycle of a product, inputs and outputs of materials must balance [29–31]. These basic concepts have been inherited and recently more precisely defined [32]. For example, building on basic ontological concepts of "object" (a collective term for products, goods, materials, resources, substances, etc.) and "event" (a collective term for accumulation, transport, transformation, dissipation, etc.), a stock is redefined as "a set of objects of interest" whereas a process is redefined as "the objects of interest that are involved in [one or several] events". Following these definitions, a flow is defined to describe an event "where objects are preserved and move from one set a to another set b" [32].

van der Voet et al. [12, 33] proposed a three-step framework for material flow analysis. First, the analyzer needs to define the system boundaries in terms of time, space (e.g., sector vs. political vs. ecological boundaries) and material (e.g., substance or substance group). The analyzer also needs to specify possible subsystems or compartments if there are any. Second, the analyzer quantifies the stocks and flows, by either simply inventorying them using a flowchart ("bookkeeping" or "accounting") or calculating them based on inputs ("modeling"). Third, the analyzer interprets the results by specifying the contributions of the material to a specific environmental issue or environmental impact. Sometimes such an interpretation involves linking the model outputs with an environmental policy.

1.3.1.1 Identification of Systems

The flow of products departing from an anthropospheric compartment (i.e., outflows) depends on the flows entering the compartment (i.e., inflows) and the property of the compartment (i.e., the efficiencies that inflows are converted to outflows). For instance, the demolition of buildings a year is a function of annual constructions of buildings in previous years (which is further determined by population and concrete density) and the average lifetime of dwelling buildings [34]. Although the relationship between inflows and outflows is never exactly linear, it is justifiable to use linear expressions in most quantifications. That is, the outflow (i.e., the mass per time) from an anthropospheric compartment, denoted as O, can be expressed as a linear function combining all past and current rates of inflows, denoted as I (t denotes the time) [35] (Eq. 1.1):

$$O(t) = \alpha(0) \cdot I(t) + \alpha(1) \cdot I(t-1) + \alpha(2) \cdot I(t-2) + \cdots + \alpha(t-1) \cdot I(1) \quad (1.1)$$

In this input-output relationship, the coefficients $\alpha(i)$, termed as "transfer coefficients" (to be discussed later), connect the inflows and outflows, hence serving as blocks for structuring the anthroposphere. In the literature, the relationship has been given different names when different sets of the coefficients are assigned. For example, if all coefficients are zero but one coefficient other than $\alpha(0)$ is 1, the relationship describes a "delay model" that has been described by van der Voet et al. [36]. If coefficients $\alpha(i) = C^{i+1}$ (C is a constant smaller than 1), the relationship describes a "leaching model" that has been described by van der Voet et al. [36]. Furthermore, if coefficients $\alpha(i) = e^{-ki}$ (k is a constant), the relationship describes a pseudo-first-order reaction.

Depending on the *temporal* relationship between inflows and outflows, an anthropospheric system belongs to one of the following three classes: time-invariant, static and dynamic. First, an anthropospheric system is called *time-invariant* if all these coefficients are constant, i.e., independent of time [35]. This means that the structure of the anthroposphere does not change with time, nor depends on the magnitude of inflows. In other words, the response of outflows to inflows is identical regardless of whether we observe it in the past, present or future; outflows do not depend on the particular time that the inflows are introduced. For example, if the use of 1 kg plastics in 1980 led to the generation of 0.5 kg plastic waste in 1989, then we can anticipate that the use of 1 kg plastics in 2020 will also lead to the generation of 0.5 kg plastic waste in 2029. Of course, we should bear in mind that the time-invariance might not always be valid because the generation of waste can decrease with time due to the increasing material utilization efficiency. However, such a time-invariant assumption is acceptable if we characterize the modeled period using a time-averaged structure of the anthroposphere. In particular, an anthropospheric system is called *steady-state* if all inflows are constant as well; in this case, outflows, and the mass of materials stored in the stage, remain unchanged with time. The main application of such a steady-state model is to evaluate the "intrinsic" property of an anthropospheric system, for example, to analyze the cause of an environmental problem by tracing back the major inflows, or to predict the effectiveness of a pollution abatement measure by projecting the magnitude of outflows [33].

Second, an anthropospheric system is called *static* if only the coefficient associated with the current inflow ($\alpha(0)$) is non-zero [35]. This means that the outflow is a function of the inflow at the same moment, irrelevant to past inflows. Inflows are cleared out immediately after entering a stage, and there is no retention, nor net accumulation, of materials in a stage. In other words, we do not consider the "lifespan" of materials. As such, a stage without net material storage is often labeled as *instantaneous* (or *transient*).

Furthermore, an anthropospheric system is determined to be *dynamic*, if both the coefficients and inflows are time-variant [35]. In this way, we can calculate not only the long-term equilibrium of the anthroposphere but also the route towards this equilibrium and the time required for equilibration. Dynamic modeling enables characterization of the long-term temporal trend in material use and waste in response to regulations or policies, in addition to the "intrinsic" property of an anthropospheric system; therefore, it is suitable for "what-if" scenario analysis [33].

1.3.1.2 Product Lifespan

As indicated above, in-use stocks can hold products for a long time, during which a product maintains its desired or required function without excessive expenditure on maintenance or repair [37]. This period is called the "product lifespan". The maximum period before a product's expiration or dysfunction is denoted as the *technical lifespan*, which is often longer than its *service lifespan*, i.e., the time interval between sale and discarding because products can be discarded before it becomes useless for social and cultural reasons. For example, we may discard a smartphone when a new generation of smartphones is marketed, even though the cell phone is still functioning well. In this book, we use the term "lifespan" to specifically refer to the service lifespan unless otherwise indicated. Even for the service lifespan, varied definitions exist. For example, the "domestic service lifespan" defines the period from product shipment to discard, regardless of single or multiple uses, whereas the "total service lifespan" includes the domestic service lifespan, plus the time for product shipment, distribution, and waste collection [38]. Oftentimes, the domestic service lifespan approximates the total service lifespan because the time for product shipment, distribution, and waste collection is short relative to the time in service. As such, we will not discriminate between the domestic and total service lifespans in this book. For a comprehensive review and summary of various definitions of lifespan and the methodologies of estimation, readers are recommended to read Oguchi et al. [39]. In addition, we remind the readers to distinguish the product lifespan from "age", which describes the time elapsed since the use of a product begins [38].

In most cases, the product lifespan varies between owners or use conditions. Therefore, rather than a fixed number, it is more appropriate to describe the lifespan using a statistical distribution (e.g., the Gaussian or Weibull distribution), which represents the probabilities (or percentages) that a product is discarded at different ages. Such a *product lifespan distribution function*, $f_{\text{product}}(\tau)$, relates the numbers of discarded products in individual observation years, $G(\tau)$, to the total number of shipped products in the shipment year (or the sale year), P, that is [39]:

$$f_{\text{product}}(\tau) = \frac{G(\tau)}{P}. \tag{1.2}$$

From a mass-balance perspective, $f_{\text{product}}(\tau)$ values in individual years should sum up to 100%. That is, the products shipped or sold in a given year will eventually completely enter the waste stream if we observe this cohort for a sufficiently long time. Using the lifespan distribution function, $f_{\text{product}}(\tau)$, we can express the number of products that are discarded and sent to the waste stream in a given observation year, $O(t)$, as a linear combination of the annual shipments or sales numbers in previous years, $I(i)$ ($i = 1, 2, \ldots, t$) [39] (Eq. 1.3):

$$O(t) = f_{\text{product}}(0) \cdot I(t) + f_{\text{product}}(1) \cdot I(t-1) + f_{\text{product}}(2) \cdot I(t-2)$$
$$+ \cdots + f_{\text{product}}(t-1) \cdot I(1). \tag{1.3}$$

Comparing Eqs. (1.1) and (1.3), we can conclude that the transfer coefficients in Eq. (1.1) are essentially the values of the lifespan distribution function $f_{\text{product}}(\tau)$ in individual years.

Based on the lifespan distribution function, we can define a *survival function* or a *reliability function*, $s_{\text{product}}(\tau)$, as the probabilities (or percentages) that a product remains in service at different ages, which is a complementary cumulative distribution function (or called as a tail distribution) of the lifespan distribution function $f_{\text{product}}(\tau)$:

$$s_{\text{product}}(\tau) = 1 - \sum_{i=0}^{\tau} f_{\text{product}}(i). \tag{1.4}$$

The product lifespan can be very long for "articles" (here we tentatively adopt the REACH definition) such as buildings and construction materials; it can also be short for others such as plastic packages. Likewise, we can view "products" (again, we tentatively adopt the REACH definition), e.g., pesticides, cosmetics and personal care products, as special variants of "articles" whose lifespans are "instantaneous", although we recognize this view may not be intuitively comprehensible. For instance, the "lifespan" of a pesticide is assumed to terminate once pests are controlled; in other words, excess pesticide residuals remaining in soil or vegetation after application can be seen as "waste" [40, 41].

Note that the lifespan distribution function and survival function can change with time. A typical example is that the lifespans of electronic devices are becoming shorter and shorter because of the accelerating technical advancements (e.g., as described by Moore's law in the field of integrated circuit chips). For instance, Babbitt et al. [42] observed that the lifespan of personal computers has declined from a mean of 10.7 years in 1985 to 5.5 years in 2000, based on statistics among 29,000 personal computers purchased by a university during this period. The lifespan distribution function and survival function also vary between regions, depending on the regional economic level and consumer behavior. For example, Oguchi and Fuse [43] observed that the average lifespan of passenger cars varies substantially among 17 countries, ranging from 13.0 years in South Korea and Ireland to 22.6 years in Australia. In Chap. 5 of this book, we will consider the variation in the lifespan distribution function of buildings in different regions.

1.3.1.3 Transfer Coefficients

From Eq. (1.1) we can see that the transfer coefficients denote the partitioning of a material in different processes. In general, a transfer coefficient, TC_{ij}, defines the fraction of an *inflow i* that ends up in a certain outflow *j* [22]. In some particular cases, the transfer coefficient can also represent the faction of a *stock* that departs from the system. The combination of all transfer coefficients determines the structure of an anthropospheric system.

Usually, transfer coefficients have various names in different specific cases or for different specific purposes. In the literature, transfer coefficients appeared as leaching or emission factors in quantification of the emission of a substance to the environment (e.g., Eckelman and Graedel [44]), as waste factors, or fractionation ratios, in quantification of the generation of leftover materials from an industrial process (e.g., Elshkaki et al. [45]), or as loss rates in quantification of the dissipation of a substance from the anthroposphere (e.g., Chen et al. [46]). In this book, emission factors, waste factors and decomposition factors that we use for calculations are examples of transfer coefficients (see definitions in Chap. 2).

It is worthwhile to emphasize that, transfer coefficients are not necessarily constant in time because they are dependent on not only materials per se but also the ever-advancing technologies involved in a process [28, 31]. However, they can be regarded as constants within a certain temporal range if necessary.

1.3.2 Modeling the Fate of Chemicals in the Environment

Environmental chemists use multimedia mass-balance models [47] to describe the fate of chemicals in the physical environment. Many CiPs, especially the troublesome ones like PCBs and PBDEs, show a strong tendency to partition between multiple environmental media; we thus need to take into account their multimedia behavior. Typical multimedia mass-balance models often comprise three major components: (i) emission information, which is the starting point of a simulation, (ii) chemical properties (e.g., partition coefficients and irreversible reactivity), and (iii) environmental parameters (e.g., atmospheric and oceanic flow rates), which characterize the basic conditions of an environment.

The fugacity-based approach [20], and its "sibling" the activity-based approach [48], have been mainstreamed in environmental transport and fate modeling. In traditional chemical thermodynamic theory, a fugacity defines the tendency that a molecular species (e.g., a CiP in our case) escapes from the phase in which it resides to an adjacent one [49]. A molecular species spontaneously departs from a phase with a higher fugacity and enters a phase with a lower fugacity, which lowers the fugacity of the former phase but elevates the latter. The interphase migration of molecular species continues until the fugacities of the two phases equate each other, i.e., a status that the system attains a thermodynamic equilibrium [49]. In this sense, the fugacity (i) measures to what extent a system deviates from the thermodynamic equilibrium, and (ii) determines the direction of diffusion of a molecular species. The fugacity is expressed in a unit of partial pressure.

Since 1979, Don Mackay and collaborators [50–52] have adopted such a microscopic concept to characterize macroscopic transport of chemicals between environment media. They postulate that the fugacity defines the tendency that a chemical departs from an environmental compartment to enter another [20]. The concentration (C, in mol m^{-3}), representing the amount of a chemical present in an environmental compartment, is a product of the fugacity capacity (Z, in mol m^{-3} Pa^{-1}), representing the capability or affinity that an environmental compartment holds the chemical, and the fugacity (f, in Pa):

$$C = Z \cdot f. \tag{1.5}$$

The Z-value of an environmental compartment (or referred to a "bulk" compartment) depends on its phase composition: it is an average of Z-values of individual immiscible phases (e.g., air, water, and organic matter) weighted by their volume fractions in this environmental compartment. For instance, we can expect the Z-values to differ between sandy (with a low organic matter content) and fertile soils (with a high organic matter content). In this way, the Z-value incorporates the properties of an environmental compartment into the calculation. For a given chemical, the ratio of the Z-values of a pair of immiscible phases (subscripts 1 and 2) is equal to the "partition coefficient" (to be discussed later) between them:

$$K_{12} = \frac{Z_1}{Z_2}. \tag{1.6}$$

Equation (1.6) indicates that the Z-value also reflects the properties of a chemical. Since the partition coefficients are temperature dependent, Z-values also depend on the temperature of environment.

The rate of an environmental process that leads to a mass change in an environmental medium (N, in mol h^{-1}), e.g., chemical reactions, advective transfer in a moving medium, and diffusive exchange between two environmental media, is a product of the D-value (in mol Pa^{-1} h^{-1}) and the fugacity,

$$N = D \cdot f. \tag{1.7}$$

The calculation of D-value differs between particular processes. For example, to calculate the D-value of an irreversible chemical reaction in an environmental compartment, we need to know an irreversible reaction rate constant (k), a volume of the medium, and a Z-value of the medium. Since it is both the chemical itself and the environment in which it resides that determine the irreversible reaction rate constant, D-values reflect the combination of properties of the chemical and environment.

By integrating the Z-values of individual environmental media and the D-values of individual processes, we can calculate the mass balance of a chemical in an environment.

There is now a wide acceptance that the variability in the environmental fate of chemicals is primarily attributable to the variability in physicochemical properties of chemicals (i.e., partition coefficients and irreversible reactivity), rather than that in properties of the environment (e.g., temperature and wind speed) [53]. This is because physicochemical properties can differ between chemicals by up to 12 orders of magnitude, whereas the properties of the environment often differ by a factor 10 to 20 only [48].

1.3.2.1 Identification of Systems

In the multimedia mass-balance modeling, the described environment can be either "evaluative" or "real" [54]. The former represents a hypothetical, fictitious environment, which generalizes and abstracts typical characteristics of the actual environment in a region. The aim of using an evaluative model is not to reproduce observed concentrations and fluxes, but to find out some general features regarding the chemical fate (e.g., major pathways and compartmental distribution), or screen and prioritize chemicals based on assessment metrics. As such, the evaluation of the model's performance is not straightforward because direct observations from some specific real environments may not always be representative of the generic situation [55]. Examples of evaluative models include the EQC [56] and Simple-Box [57] models. In contrast, a "real" model describes a particular site-specific, but well-defined, situation; evaluation of such a model is generally more feasible by comparing the model output with observations [55]. The "real" models are often spatially resolved; some of them are even temporally resolved. For example, a model named Globo-POP [58] divides the real global environment into ten interconnected climate zones and output time-dependent (usually monthly resolved) results, whereas another named CliMoChem [59] divides the real global environment into 10–120 latitudinal zones and output time-dependent (usually monthly resolved) results. In general, the equations that are used in both evaluative and real models are the same; only the environmental parameters are different because of different application purposes.

A multimedia mass-balance model can involve a dimension of time in its calculation. Models that generate time-dependent results are labeled as "Level IV" [20], in which either, or both, of the structure of the modeled environment (e.g., D-values) and inputs (e.g., emissions) is time-dependent. Note that some literature (e.g., Hertwich [60]) restricted the Level IV models to those with a fixed structure of the modeled environment but time-variant inputs, but this is not a requirement originally stated in Don Mackay's work. In fact, a number of models that consider seasonal [58, 59] and inter-annual [61] variations in environmental conditions (e.g., temperature, precipitation rate, and wind speed) have also been classified as Level IV. In contrast, models that generate time-independent results are either Level I, II, or III [20], which do not consider the temporal change in environmental contamination. A model describing the chemical equilibrium distribution in a closed system (no input, no removal) is classified as Level I, a model describing the chemical equilibrium distribution in an open system (with input, with removal) as Level II, and a model requiring no equilibrium assumptions is classified as Level III [20]. In scientific research, the Level III and IV models are most common.

1.3.2.2 Partition Coefficients

A partition coefficient, i.e., "the [ratio] of concentrations between two immiscible phases" [62], indicates the phase that a chemical prefers to stay, thus being a key factor determining the multimedia mass distribution and the fate of a chemical in the physical environment. Routinely used pairs of unitless partition coefficients include those between air and water (K_{AW}), between n-octanol and water (K_{OW}), and between n-octanol and air (K_{OA}). The three phases have been chosen, because they can be readily obtained in a manner of pure phases, and hence, the measurements are usually reliably reproducible. The three partition coefficients also constitute a "thermodynamic triangle", which enables one partition coefficient to be derived from the other two [63]. Here, n-octanol is used as a surrogate solvent to mimic the organic phase (e.g., lipid) because it has a ratio of carbon, hydrogen, and oxygen atoms similar to that present in natural lipids [20]. A chemical is lipophilic if it tends to partition into the octanol phase and lipophobic if otherwise. In a few studies, the Henry's Law constant in Pa m^3 mol^{-1} is used; it can be converted from, and thus equivalent to, the unitless K_{AW}. Other phases, e.g., mineral phases, are generally not considered in characterization of phase partitioning because of their negligible roles in holding organic chemicals except in certain extremely dry conditions. For more information on the concept and evolution of partition coefficients, readers are recommended to refer to Mackay et al. [62].

For chemicals dissociable in water, partition coefficients can be defined for both neutral molecules and their dissociated ions, which often substantially distinct from each other. The partition coefficient of either a non-ionized or an ionized species is denoted using the letter "K", whereas an overall partition coefficient of the totality of both of them is expressed as "D" (sometimes called "effective K") [64]. Note that the partition coefficient D is different from the D-value mentioned above. D is an average of Ks of individual species, weighted by their relative abundance in a solution; as such, it is dependent on the pH of the solution and pK_a of the molecule [64–66]. In Chap. 4, we will convert Ks to Ds when calculating the environmental fates of PFCAs.

1.3.2.3 Irreversible Reaction Half-Lives

Another key factor governing the multimedia distribution and fate of a chemical is the irreversible reaction in the environment. The irreversible reaction leads to a decline in the chemical mass in the environment. When the decline can be fitted with an exponential function, the slope of this fit provides a "half-life" (HL), i.e., the length of time required for reducing the chemical mass to half of its the initial level. The irreversible reaction half-life can be derived from the corresponding rate constant (k) mentioned earlier using Eq. (1.8),

$$HL = \frac{\ln 2}{k}. \tag{1.8}$$

A chemical is labile if it has short irreversible reaction half-lives and recalcitrant if otherwise. The irreversible reaction of chemicals in air occurs mainly through hydroxylation with OH radicals in the gaseous phase. The reaction between chemicals in the particulate phase with OH radicals can also be important in some cases. The reactions in various environmental surface compartments (e.g., soil, sediment, and surface water) are normally interdependent because different surface compartments often share similar hydrolytic, oxidative, and microbial degradation processes. That is, if a chemical is recalcitrant in the surface water, it is also likely to be recalcitrant in soil and sediment. When no chemical-specific measured data are available, as a rule of thumb, modelers often assume the half-life in surface water as a "reference" value, and the half-life in soil and sediment to be twice and ten times the reference value, respectively [67, 68]. The EU Technical Guidance Document further assumes that the half-life in soil is 20 times the reference value if the partition coefficient $K_{soil-water}$ is great than 100 L kg^{-1}, and 200 times the reference value if $K_{soil-water}$ is great than 1000 L kg^{-1} [69].

1.4 Research Needs and Objectives

1.4.1 Need for an Integrative Modeling Framework

As indicated above, the CiP issues are being addressed by scientists from two independent disciplines (industrial ecologists vs. environmental chemists) using two separate approaches (material flow analysis vs. multimedia mass-balance modeling) from two parallel perspectives (an industrial or human angle vs. an environmental or non-human angle). Such a disunited research paradigm contradicts the systematic nature of the anthroposphere and physical environment because the two subsystems are inherently interconnected and interdependent. The disunion also creates problems for both research communities.

For example, industrial ecologists are often plagued with the fact that the estimated anthrospheric fate of products is not directly observable, and hence not readily verifiable in practice. A common case is that we can hardly measure the emissions of a substance or the waste flow of a product by first-hand observation. While the in-use stock of a product sometimes can be roughly estimated by summing up the numbers used in individual applications, the estimate is rather uncertain because the survey usually cannot be exhaustive, i.e., not a census. Therefore, estimates by different industrial ecologists can often be quite discrepant because of inconsistent boundary definitions, data sources, product categories considered, etc. What is worse is that we can seldom tell which is more realistic. An example is a substantial variation in estimates of the in-use stock of steel in Japan, ranging from a lower end of 1.5 to 14 tonnes per capita, as summarized by Pauliuk et al. [70]. On the other hand, environmental chemists often face the dearth of emission estimates required for driving their environmental fate models. As argued by Breivik et al. [71], emission estimates "remain the least understood part of the research on overall distribution and fate of [...] chemicals in the environment". An ideal emission estimate that is well-suited for environmental fate modeling should (i) have high-resolution temporal and spatial coverage, (ii) include multiple lifecycle stages and multiple receiving environmental compartments, and (iii) consider speciation if there are multiple species (e.g., congeners or isomers) [71]. However, such favorable emission estimates are absent for most CiPs, except for some well-studied, data-rich chemicals such as PCBs [72–74], PFCAs [75] and PFSAs [76], and SCCPs [77]. Even for these CiPs, the emission estimates are incomplete or inadequate [78]. For example, the emission estimates of PCBs [72–74] fail to include the releases of PCBs from industrial processes. When applying these emission estimates to simulate the environmental fate of and human exposure to PCBs, Breivik et al. [79] and Wood et al. [80] realized that the peaking time of modeled contamination lags a decade behind the observation.

For the above reasons, it would be helpful and critical to address the CiP problems if we could integrate the modeling of the anthropospheric and environmental fates to establish a modeling continuum from chemical production to environmental concentrations. If a given in-use or waste stock is associated with a certain amount of chemical emissions that lead to a certain level of contamination in the environment, then we can evaluate whether the estimates of emissions and stock are reasonable by comparing the modeled contamination with observations. In this case, emission estimates are outputs of the anthropospheric fate modeling but required as inputs to the environmental fate modeling, thus bridging the anthroposphere and physical environment in the model representation. We need to assimilate fragmented knowledge from multidisciplinary sources to develop a holistic, comprehensive understanding of the interactions between the human socioeconomic systems and the non-human natural ecosystems. In addition to providing the observable, verifiable modeling results, the fusion of anthropospheric and environmental fate modeling also contributes to the comprehensiveness of environmental analysis and helps avoid overlooking relevant factors or unintended consequences.

1.4.2 Need for a Temporally Resolved Modeling Framework

Models without the temporal dimension are currently popular in material flow analysis and multimedia mass-balance modeling because they are computationally simple. These models involve steady-state material flow analysis and the Level I through

Level III multimedia mass-balance models, as elaborated above. The outputs from the models are steady-state, or stationary, estimates of flows or stocks. The omission of the temporal dimension during a given simulation period is tenable only in cases that:

(i) if there is an ongoing input (e.g., annual production or emission) into the model, the input is, or very close to, time-independent, and the time that a material resides in a lifecycle stage or a chemical resides in the environment is not too long; or

(ii) if the input is already absent, the input ended a considerable time ago, and the time that a material resides in a lifecycle stage or a chemical resides in the environment is sufficiently long.

These two conditions require the "age" of a material or a chemical to be irrelevant in the two systems. That is, the material/chemical entering the systems in earlier and later times are indistinguishable and have the same possibility of departing from the system.

For example, in the anthroposphere the discard flow of a type of products reaches the steady state if the market remains stable or saturated, i.e., new items are manufactured solely to replace broken, worn-out ones, and the product lifespan is relatively short (shorter than the period of observation) [36]. Examples include electronic and electric equipment, jeans, and other household goods [36]. Likewise, in the physical environment, the concentration of a chemical remains steady state if the chemical emission has ceased and the chemical is recalcitrant [55]. An example is environmental contamination long after a ban on the agricultural use of highly persistent organochlorine pesticides [81]. It is clear that many CiPs do not necessarily meet these conditions because of skyrocketing market demand, the use in long-lived consumer goods (e.g., building materials), and/or high chemical degradability in the environment. Their occurrence in the anthroposphere and environment can be highly dynamic, which, therefore, warrants temporally resolved models.

Furthermore, a temporally resolved modeling framework is necessary if predicting the temporal evolution of the anthropospheric and environmental fate is the purpose of evaluation. For example, regulators may wonder, to what extent, and how fast, the contamination of a chemical would decrease if its production and new uses are restricted or banned as per regulatory policies. The dimension of time is needed for a "what-if" scenario analysis.

1.4.3 Need for a Mechanistic Modeling Framework

Most existing studies are on a product-by-product or chemical-by-chemical basis. Of course, investigating products or chemicals one by one is time-consuming and costly, which is far from feasible when faced with the fact that millions of commercial CiPs are awaiting evaluation. In addition, the case studies rely intensively upon empirical or semi-empirical relationships without a mechanistic understanding of physical, chemical, biological and social processes involved. The empirical or semi-empirical relationships are simple approximations for the complex anthropospheric and environmental systems, whereby we can make reasonable predictions in a computationally cheap manner. For example, we can roughly derivate the number of in-use products in a region by a simple multiplicative relationship between the average product lifespan and a surveyed average annual sales [82], which are two readily available parameters. Note that although statistical techniques, e.g., the linear regression, are substantially involved in creating empirical or semi-empirical relationships, "statistics" is never a synonym of empirical or semi-empirical relationships [83]. Mechanistic simulations also involve a considerable amount of statistical techniques, e.g., stochastic models.

The disadvantages of empirical or semi-empirical relationships are also obvious. For instance, since the relationships are observation-driven, in most cases we can characterize the link between observations but cannot explain the causations behind the link. For instance, Li et al. [84] derived an empirical relationship correlating the emission density (ED, in mg km^{-2} d^{-1}) of PFOA from Chinese wastewater treatment plants with area-specific population density (PD, in cap km^{-2}) and per capita disposable income ($PCDI$, CNY cap^{-1}) of the administrative district in which the wastewater treatment plant is located:

$$ED = 6.27 \times 10^{-11} \times PD^{0.901} \times PCDI^{1.897} \left(N = 41; R^2 = 0.6660\right) \qquad (1.9)$$

However, the regression coefficients are totally different for the case of emissions of PFOS [85] from Chinese wastewater treatment plants:

$$ED = 1.25 \times 10^{-11} \times PD^{1.317} \times PCDI^{1.664} \left(N = 37; R^2 = 0.6800\right) \qquad (1.10)$$

It is close to impossible to explain such a discrepancy because the regression itself provides almost no information on how these regression coefficients are linked to the chemicals' physicochemical properties or the manners that the two chemicals are used. In addition, the empirical or semi-empirical relationships are often product- or chemical-specific, which means that they apply to only a narrow group of products or chemicals based on which the relationships were established. Their applicability domains, i.e., the conditions that these relationships can be reliably applicable, remain largely unknown. Therefore, we should be extremely cautious when applying the empirical or semi-empirical relationships to new products and chemicals and interpreting the modeling results.

In contrast, process-based mechanistic models, despite being more complicated and demanding, provide a more systematic and interpretable insight on the behavior and fate of CiPs in the anthroposphere and environment. With process-based mechanistic models, we can quantify by what means external conditions influence the individual physical, chemical, biological and social processes in the anthroposphere and environment, and in what condition a process becomes dominant. We are also able to improve the predictivity of models.

1.4.4 Research Objectives

To address the research needs above, we seek to develop an integrative modeling system that coherently couples the anthroposphere and physical environment, based on time-dependent mechanistic descriptions of physical, chemical, biological and social processes of CiPs in the two interrelated, interacting and interdependent systems. The developed modeling system will be evaluated and applied to quantify the accumulation, transport and dissipation of CiPs in the anthroposphere and physical environment, and how these processes are influenced by the properties of products/chemicals and human regulatory activities.

1.5 The Structure of the Book

The structure of this book is organized as follows:

Chapter 2 introduces the rationale and structure of the anthropospheric fate model CiP-CAFE built for this book, and its connection with an existing environmental fate model BETR-Global to create an integrated, temporally resolved and mechanistic modeling framework.

With the developed modeling framework, Chap. 3 tracks, using the case of PCBs, the major route by which CiPs disperse in the global "total" environment. Chapter 4 elucidates the temporal evolution of the contributions of side-chain fluorotelomer-based polymers degradation to the long-term occurrence of fluorotelomer alcohols and perfluoroalkyl carboxylates. Chapter 5 explains why, in the environment, we can observe a composition of CiPs that is completely different from that in the technical mixture at the production stage, using the case of three HBCDD diastereomers. Chapter 6 illustrates how a "what-if" scenario analysis is conceptualized and helps figure out the best way of minimizing emissions if different end-of-life management strategies are taken in the future.

References

1. UNEP (2019) Global chemicals outlook II. From legacies to innovative solutions: implementing the 2030 agenda for sustainable development. United Nations Environment Programme, Nairobi
2. Alaee M, Arias P, Sjödin A, Bergman Å (2003) An overview of commercially used brominated flame retardants, their applications, their use patterns in different countries/regions and possible modes of release. Environ Int 29(6):683–689
3. Covaci A, Harrad S, Abdallah MAE, Ali N, Law RJ, Herzke D, de Wit CA (2011) Novel brominated flame retardants: a review of their analysis, environmental fate and behaviour. Environ Int 37(2):532–556
4. UNEP (2012) Global chemicals outlook I. Towards sound management of chemicals. United Nations Environment Programme, Nairobi

5. Attina TM, Hauser R, Sathyanarayana S, Hunt PA, Bourguignon J-P, Myers JP, DiGangi J, Zoeller RT, Trasande L (2016) Exposure to endocrine-disrupting chemicals in the USA: a population-based disease burden and cost analysis. Lancet Diabetes Endocrinol 4(12):996–1003
6. Trasande L, Myers JP, DiGangi J, Bellanger M, Legler J, Skakkebaek NE, Heindel JJ, Zoeller RT, Hass U, Kortenkamp A, Hauser R, Grandjean P (2015) Estimating burden and disease costs of exposure to endocrine-disrupting chemicals in the European Union. J Clin Endocrinol Metab 100(4):1245–1255
7. SAICM (2006) Overarching policy strategy of the strategic approach to international chemicals management (SAICM)
8. UNEP (2015) The chemicals in products programme. United Nations Environment Programme, Geneva
9. European Union (2006) Registration, evaluation, authorisation and restriction of chemicals (REACH). Regulation (EC) No 1907/2006
10. ECHA (2017) EU guidance on requirements for substances in articles (Version 4.0). European Chemicals Agency (ECHA), Helsinki
11. Tibben-Lembke RS (2002) Life after death: reverse logistics and the product life cycle. Int J Phys Distrib Logist Manag 32(3):223–244
12. de Haes HU, van der Voet E, Kleijn R (1997) Substance flow analysis (SFA), an analytical tool for integrated chain management. In: Bringezu S, Fischer-Kowalski M, Kleijn R, Palm V (eds) Regional and national material flow accounting: from paradigm to sustainability, Proceedings of the ConAccount workshop, Leiden, The Netherlands, pp 32–42
13. Manahan SE (2000) Environmental chemistry, 8th edn. CRC Press, Boca Ration, FL
14. Schellnhuber HJ (1999) "Earth system" analysis and the second Copernican revolution. Nature 402(6761):C19–C23
15. Eduljee GH (2001) Budget and source inventories. In: Harrad S (ed) Persistent organic pollutants: environmental behaviour and pathways of human exposure. Springer, US, Boston, MA, pp 1–28
16. Harrad SJ, Sewart AP, Alcock R, Boumphrey R, Burnett V, Duarte-Davidson R, Halsall C, Sanders G, Waterhouse K, Wild SR, Jones KC (1994) Polychlorinated biphenyls (PCBs) in the British environment: sinks, sources and temporal trends. Environ Pollut 85(2):131–146
17. Abbasi G, Buser AM, Soehl A, Murray MW, Diamond ML (2015) Stocks and flows of PBDEs in products from use to waste in the US and Canada from 1970 to 2020. Environ Sci Technol 49(3):1521–1528
18. Wania F, Su Y (2004) Quantifying the global fractionation of polychlorinated biphenyls. Ambio 33(3):161–168
19. Wania F (2007) A global mass balance analysis of the source of perfluorocarboxylic acids in the Arctic Ocean. Environ Sci Technol 41(13):4529–4535
20. Mackay D (2001) Multimedia environmental models: the fugacity approach, 2nd edn. CRC Press, Boca Raton, FL
21. Li L, Wania F (2018) Elucidating the variability in the hexabromocyclododecane diastereomer profile in the global environment. Environ Sci Technol 52(18):10532–10542
22. Baccini P, Brunner PH (2012) Metabolism of the anthroposphere: analysis, evaluation, design, 2nd edn. The MIT Press, Cambridge, MA and London
23. Bringezu S, Moriguchi Y (2002) Material flow analysis. In: Ayres RU, Ayres LW (eds) A handbook of industrial ecology. Edward Elgar, Cheltenham, UK
24. Binder C, Bader H-P, Scheidegger R, Baccini P (2001) Dynamic models for managing durables using a stratified approach: the case of Tunja, Colombia. Ecol Econ 38(2):191–207
25. Müller E, Hilty LM, Widmer R, Schluep M, Faulstich M (2014) Modeling metal stocks and flows: a review of dynamic material flow analysis methods. Environ Sci Technol 48(4):2102–2113
26. Chen W, Graedel TE (2012) Anthropogenic cycles of the elements: a critical review. Environ Sci Technol 46(16):8574–8586

27. Keller AA, McFerran S, Lazareva A, Suh S (2013) Global life cycle releases of engineered nanomaterials. J Nanoparticle Res 15(6):1692
28. Brunner PH, Rechberger H (2004) Practical handbook of material flow analysis. CRC Press LLC, Boca Raton, Florida
29. Ayres RU, Kneese AV (1969) Production, consumption, and externalities. Am Econ Rev 59(3):282–297
30. Ayres RU (1978) Resources, environment and economics: applications of the materials/energy balance principle. Wiley and Sons, New York
31. Ayres RU (1989) Industrial metabolism. In: Ausubel JH, Sladovich HE (eds) Technology and Environment. National Academy Press, Washington, DC
32. Pauliuk S, Majeau-Bettez G, Müller DB, Hertwich EG (2015) Toward a practical ontology for socioeconomic metabolism. J Ind Ecol 20(6):1260–1272
33. van der Voet E, Heijungs R, Mulder P, Huele R, Kleijn R, van Oers L (1995) Substance flows through the economy and environment of a region. Environ Sci Pollut Res 2(3):137–144
34. Müller DB (2006) Stock dynamics for forecasting material flows—case study for housing in The Netherlands. Ecol Econ 59(1):142–156
35. Bauer G, Deistler M, Gleiß A, Glenck E, Matyus T (1997) Identification of material flow systems. Environ Sci Pollut Res 4(2):105
36. van der Voet E, Kleijn R, Huele R, Ishikawa M, Verkuijlen E (2002) Predicting future emissions based on characteristics of stocks. Ecol Econ 41(2):223–234
37. Cooper T (1994) Beyond recycling: the longer life option. New Economics Foundation, London
38. Murakami S, Oguchi M, Tasaki T, Daigo I, Hashimoto S (2010) Lifespan of commodities, part I: the creation of a database and its review. J Ind Ecol 14(4):598–612
39. Oguchi M, Murakami S, Tasaki T, Daigo I, Hashimoto S (2010) Lifespan of commodities, part II: methodologies for estimating lifespan distribution of commodities. J Ind Ecol 14(4):613–626
40. Li Y-F, Scholtz M, Van Heyst B (2000) Global gridded emission inventories of α-hexachlorocyclohexane. J Geophys Res Atmos 105(D5):6621–6632
41. Li Y-F, Scholtz MT, Van Heyst BJ (2003) Global gridded emission inventories of β-hexachlorocyclohexane. Environ Sci Technol 37(16):3493–3498
42. Babbitt CW, Kahhat R, Williams E, Babbitt GA (2009) Evolution of product lifespan and implications for environmental assessment and management: a case study of personal computers in higher education. Environ Sci Technol 43(13):5106–5112
43. Oguchi M, Fuse M (2015) Regional and longitudinal estimation of product lifespan distribution: a case study for automobiles and a simplified estimation method. Environ Sci Technol 49(3):1738–1743
44. Eckelman MJ, Graedel T (2007) Silver emissions and their environmental impacts: a multilevel assessment. Environ Sci Technol 41(17):6283–6289
45. Elshkaki A, Van der Voet E, Van Holderbeke M, Timmermans V (2004) The environmental and economic consequences of the developments of lead stocks in the Dutch economic system. Resour Conserv Recycl 42(2):133–154
46. Chen W, Shi L, Qian Y (2010) Substance flow analysis of aluminium in mainland China for 2001, 2004 and 2007: exploring its initial sources, eventual sinks and the pathways linking them. Resour Conserv Recycl 54(9):557–570
47. MacLeod M, Scheringer M, McKone TE, Hungerbuhler K (2010) The state of multimedia mass-balance modeling in environmental science and decision-making. Environ Sci Technol 44(22):8360–8364
48. Mackay D, Arnot JA (2011) The application of fugacity and activity to simulating the environmental fate of organic contaminants. J Chem Eng Data 56(4):1348–1355
49. Lewis GN (1901) The law of physico-chemical change. Proc Am Acad Arts Sci 37(3):49–69
50. Mackay D (1979) Finding fugacity feasible. Environ Sci Technol 13(10):1218–1223
51. Mackay D, Paterson S (1981) Calculating fugacity. Environ Sci Technol 15(9):1006–1014
52. Mackay D, Paterson S (1982) Fugacity revisited. Environ Sci Technol 16(12):654A–660A
53. Webster E, Mackay D, Di Guardo A, Kane D, Woodfine D (2004) Regional differences in chemical fate model outcome. Chemosphere 55(10):1361–1376

54. Mackay D, Webster E, Cousins I, Cahill T, Foster K, Gouin T (2001) An introduction to multimedia models (CEMC Report No. 200102). Canadian Environmental Modelling Centre, Peterborough
55. Wania F, Mackay D (1999) The evolution of mass balance models of persistent organic pollutant fate in the environment. Environ Pollut 100(1–3):223–240
56. Mackay D, Di Guardo A, Paterson S, Cowan CE (1996) Evaluating the environmental fate of a variety of types of chemicals using the EQC model. Environ Toxicol Chem 15(9):1627–1637
57. van de Meent D (1993) SIMPLEBOX: a generic multimedia fate evaluation model (RIVM Report 672720001). Rijksinstituut voor Volksgezondheid en Milieu (RIVM), Bilthoven
58. Wania F, Mackay D (1995) A global distribution model for persistent organic chemicals. Sci Total Environ 160:211–232
59. Scheringer M, Wegmann F, Fenner K, Hungerbühler K (2000) Investigation of the cold condensation of persistent organic pollutants with a global multimedia fate model. Environ Sci Technol 34(9):1842–1850
60. Hertwich EG (2001) Fugacity superposition: a new approach to dynamic multimedia fate modeling. Chemosphere 44(4):843–853
61. Wöhrnschimmel H, MacLeod M, Hungerbuhler K (2013) Emissions, fate and transport of persistent organic pollutants to the Arctic in a changing global climate. Environ Sci Technol 47(5):2323–2330
62. Mackay D, Celsie AK, Parnis JM (2015) The evolution and future of environmental partition coefficients. Environ Rev 24(1):101–113
63. Cole JG, Mackay D (2000) Correlating environmental partitioning properties of organic compounds: the three solubility approach. Environ Toxicol Chem 19(2):265–270
64. Schwarzenbach RP, Gschwend PM, Imboden DM (2003) Environmental organic chemistry, 2nd edn. Wiley & Sons Inc, Hoboken, NJ
65. Schellenberg K, Leuenberger C, Schwarzenbach RP (1984) Sorption of chlorinated phenols by natural sediments and aquifer materials. Environ Sci Technol 18(9):652–657
66. Jafvert CT, Westall JC, Grieder E, Schwarzenbach RP (1990) Distribution of hydrophobic ionogenic organic compounds between octanol and water: organic acids. Environ Sci Technol 24(12):1795–1803
67. Fenner K, Scheringer M, MacLeod M, Matthies M, McKone T, Stroebe M, Beyer A, Bonnell M, Le Gall AC, Klasmeier J (2005) Comparing estimates of persistence and long-range transport potential among multimedia models. Environ Sci Technol 39(7):1932–1942
68. Wania F (2006) Potential of degradable organic chemicals for absolute and relative enrichment in the Arctic. Environ Sci Technol 40(2):569–577
69. ECB (2003) Technical guidance document on risk assessment. European Chemicals Bureau (ECB), Italy
70. Pauliuk S, Wang T, Müller DB (2013) Steel all over the world: estimating in-use stocks of iron for 200 countries. Resour Conserv Recycl 71:22–30
71. Breivik K, Vestreng V, Rozovskaya O, Pacyna JM (2006) Atmospheric emissions of some POPs in Europe: a discussion of existing inventories and data needs. Environ Sci Policy 9(7):663–674
72. Breivik K, Sweetman A, Pacyna JM, Jones KC (2002) Towards a global historical emission inventory for selected PCB congeners—a mass balance approach: 1. global production and consumption. Sci Total Environ 290(1):181–198
73. Breivik K, Sweetman A, Pacyna JM, Jones KC (2002) Towards a global historical emission inventory for selected PCB congeners—a mass balance approach: 2. emissions. Sci Total Environ 290(1):199–224
74. Breivik K, Sweetman A, Pacyna JM, Jones KC (2007) Towards a global historical emission inventory for selected PCB congeners—a mass balance approach: 3. an update. Sci Total Environ 377(2):296–307
75. Wang Z, Cousins IT, Scheringer M, Buck RC, Hungerbühler K (2014) Global emission inventories for C4–C14 perfluoroalkyl carboxylic acid (PFCA) homologues from 1951 to 2030, Part I: production and emissions from quantifiable sources. Environ Int 70:62–75

76. Wang Z, Boucher JM, Scheringer M, Cousins IT, Hungerbühler K (2017) Toward a comprehensive global emission inventory of C4–C10 perfluoroalkanesulfonic acids (PFSAs) and related precursors: focus on the life cycle of C8-based products and ongoing industrial transition. Environ Sci Technol 51(8):4482–4493

77. Glüge J, Wang Z, Bogdal C, Scheringer M, Hungerbühler K (2016) Global production, use, and emission volumes of short-chain chlorinated paraffins—a minimum scenario. Sci Total Environ 573:1132–1146

78. Breivik K, Alcock R (2002) Emission impossible? The challenge of quantifying sources and releases of POPs into the environment. Environ Int 28(3):137–138

79. Breivik K, Czub G, McLachlan MS, Wania F (2010) Towards an understanding of the link between environmental emissions and human body burdens of PCBs using CoZMoMAN. Environ Int 36(1):85–91

80. Wood SA, Armitage JM, Binnington MJ, Wania F (2016) Deterministic modeling of the exposure of individual participants in the National Health and Nutrition Examination Survey (NHANES) to polychlorinated biphenyls. Environ Sci Process Impacts 18(9):1157–1168

81. Cao H, Tao S, Xu F, Coveney RM, Cao J, Li B, Liu W, Wang X, Hu J, Shen W (2004) Multimedia fate model for hexachlorocyclohexane in Tianjin, China. Environ Sci Technol 38(7):2126–2132

82. Oguchi M, Daigo I (2017) Measuring the historical change in the actual lifetimes of consumer durables. In: Bakker C, Mugge R (eds) Product lifetimes and the environment 2017-conference proceedings. Delft University of Technology and IOS Press, pp 319–323

83. Thakur AK (1991) Model: mechanistic vs empirical. In: Rescigno A, Thakur AK (eds) New trends in pharmacokinetics. Springer, US, Boston, MA, pp 41–51

84. Li L, Zhai Z, Liu J, Hu J (2015) Estimating industrial and domestic environmental releases of perfluorooctanoic acid and its salt in China. Chemosphere 129:100–109

85. Xie S, Lu Y, Wang T, Liu S, Jones K, Sweetman A (2013) Estimation of PFOS emission from domestic sources in the eastern coastal region of China. Environ Int 59:336–343

Chapter 2
Developing Models for Tracking the Fate of Chemicals in Products in the Total Environment

Abstract Dynamically tracking flows and stocks of chemicals in products (CiPs) in the global anthroposphere and physical environment is essential for understanding their environmental fates and risks. The complex behavior of CiPs during production, industrial activities, use and waste disposal makes this task particularly challenging. In this chapter, we develop a global anthropospheric fate model, named Chemicals in Products—Comprehensive Anthropospheric Fate Estimation (CiP-CAFE). Provided with an array of product and chemical property parameters, the CiP-CAFE model facilitates the quantification of time-variant flows and stocks of CiPs within and between seven interconnected world regions and the generation of global-scale emission estimates. We also briefly introduce its connection with a previously published global environmental fate model, named the Berkeley-Trent Global (BETR-Global); the fusion of two models enables simulating temporally and spatially resolved masses and concentrations of chemicals.

2.1 Analogy Between Chemical Anthropospheric and Environmental Fate Modeling

The basis on which we can integrate the anthropospheric and environmental fate modeling lies in that the accumulation, transport and dissipation of mass (chemicals or products) follow similar rules in both the anthroposphere and physical environment. Such an analogy allows us to look at the human socioeconomic systems in the same manner as for non-human "natural" ecosystems.

Most fundamentally, both industrial ecologists and environmental chemists rely on the law of conservation of mass stated by Antonine Lavoisier in 1789 [1, 2]. That is, the mass can neither be created or destroyed, and it can only be rearranged or relocated between different lifecycle stages or environmental compartments in time and in space. The advocacy of the mass balance approach is not peculiar to environmental chemists [1]; it also stands out in intensive industrial ecological literature such as Ayres and Kneese [3], who argued that the failure to balance the human economic

© Springer Nature Singapore Pte Ltd. 2020   27
L. Li, *Modeling the Fate of Chemicals in Products*, Springer Theses,
https://doi.org/10.1007/978-981-15-0579-9_2

activities and environmental quality results from "viewing the production and consumption processes in a manner that is somewhat at variance with the fundamental law of the conservation of mass".

In the model representation, each lifecycle stage or environmental compartment is abstractly expressed as a "control volume"; chemicals or products can flow into and out from, and reside in, such a control volume. In a generic sense, the rate of change in mass (M) in a control volume is expressed as a difference between inflows (i.e., source processes, N_{in}) and outflows (i.e., sink processes, N_{out}), that is,

$$\frac{dM}{dt} = \sum_p N_{in} - \sum_q N_{out} \tag{2.1}$$

Here, N_{in} and N_{out} have a dimension of mass per time, measuring the rate of movement, or exchange, of chemicals between different control volumes. The task of a modeler is to mechanistically understand the individual processes and to quantitatively characterize the inflows and outflows. In this way, we can write down a series of similar differential equations, each of which represents a realistic compartment, for the anthroposphere and physical environment of interest. Outflows from one compartment can be inflows to another, by which we mathematically connect the interacted compartments. By solving these differential equations, we can obtain the estimates of chemical mass in any compartment at any given time.

We consider two categories of flows, i.e., reversible and irreversible, in our modeling. *Reversible flows* describe that chemicals move between compartments but do not leave the system. Examples in environmental fate modeling include *advection*, i.e., the transport of a chemical resulting from the movement of a bulk fluid (e.g., wind or oceanic currents) in which the chemical is present, and *diffusion*, i.e., the transport of chemicals themselves resulting from the "random walk" of chemical molecules whereas the related environmental compartments remain stationary [1]. Advection also occurs in the anthroposphere. For instance, it has been confirmed that toxic chemicals disperse worldwide along with the transboundary trade in waste electrical and electronic equipment [4]. In this case, the trade flows of waste act as carriers to "piggyback" polychlorinated biphenyls (PCBs) for the global spatial dispersal. In contrast, *irreversible flows* are driving forces for the phenomenon that chemicals permanently dissipate from the anthroposphere or physical environment and will never return. In the physical environment, for example, chemicals can disappear through chemical or biological reactions, or deposition into the deep sea [5]. An example for the anthropospheric irreversible flows is the continuous loss of fluorinated durable water repellents during the consumer use of water- and stain-proof textiles and fabrics, which causes a constant decline in water and stain repellency [6].

The analogy between the anthroposphere and physical environment also means that the fate of chemicals relies not only on the fundamental structures of systems, but also the unique properties of chemicals themselves. Environmental chemists have established that the fate of chemicals in the physical environment is mainly governed by the partition between environmental compartments and the irreversible reaction within environmental compartments [7]. Furthermore, according to the theoretical discussion in Chap. 1, it is sufficient to characterize the environmental fate of a chemical using two of the three partition coefficients K_{OW}, K_{OA}, and K_{AW} [8], and degradation half-lives in air and a "reference" surface medium [9]. These parameters are "intrinsic" properties of a chemical but can change with environmental conditions such as temperature and salinity [10]. Therefore, using condition-specific values in environmental fate modeling enables reflecting the interaction between chemicals and the physical environment. Likewise, the fate of chemicals in the anthroposphere is mainly governed by product lifespans and transfer coefficients [11]. Therefore, it is sufficient to characterize the anthropospheric fate of a chemical using lifespans of individual products in which the chemical resides, and a series of emission, waste and decomposition factors. These parameters are "intrinsic" properties of a chemical and its associated products but can change with socioeconomic conditions such as the regional economic level, contamination controlling level or consumer behavior [12]. Therefore, using condition-specific values in environmental fate modeling enables reflecting the interaction between chemicals and the physical environment.

One of the outputs from anthropospheric fate modeling is the estimates of multisource, multimedia emissions, which serve as the starting point of environmental fate modeling. Therefore, the emission flows reflect the unidirectional, irreversible movement of chemicals from the anthroposphere and physical environment, which thus bridge the anthropospheric and environmental fate modeling.

In this chapter, we will describe the development of a global anthropospheric fate model, named *Chemicals in Products—Comprehensive Anthropospheric Fate Estimation* (CiP-CAFE). Provided with an array of product and chemical property parameters, the model enables quantifying time-variant flows and stocks of CiPs within and between seven interconnected world regions and generating global scale emission estimates. We will also briefly introduce a global environmental fate model named the *Berkeley-Trent Global Model* (BETR-Global), which can be coupled with the CiP-CAFE model to calculate time-variant masses and concentrations of chemicals in seven interconnected environmental compartments across the world.

2.2 CiP-CAFE: Anthropospheric Fate Modeling

2.2.1 Structure and Conventions

The CiP-CAFE model splits the world into seven regions (Fig. 2.1a) according to (i) the development of the regional chemical industry, (ii) the status of sound management of chemicals and municipal solid waste, and (iii) major international trade origins/destinations of chemical-related commodities [13, 14]. Each region is divided into three interconnected *phases* (Fig. 2.1b), each of which further consists of a set

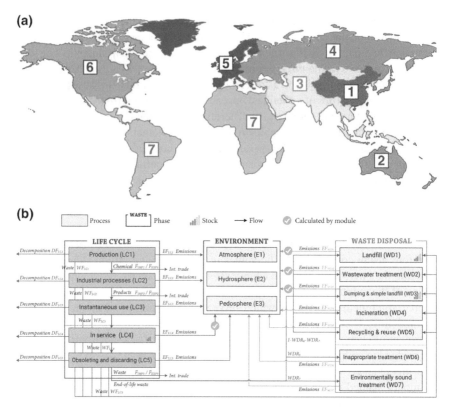

Fig. 2.1 Segmentations of (**a**) geographic regions, and (**b**) phases, processes and flows in the CiP-CAFE model

of sequential or parallel *processes* that represent events related to the CiP of interest, e.g., transformation or accumulation [15, 16].

The three phases are the:

(i) *Life Cycle* (LC) of the chemical and its associated commodities. It encompasses five sequential processes: production (LC1), industrial processes (LC2), instantaneous use (LC3), in service (LC4) and obsoleting and discarding (LC5). Such a process division applies to various CiPs as it accords with the standardized risk assessment framework for chemicals in the EU Technical Guidance Document on Risk Assessment [17]. At process LC2 and continuing to process LC5, a compound is allocated to up to five distinct parallel *applications* (APs) according to region-specific and time-variant distribution ratios; the distribution ratios of all APs sum up to 100% for every single year (Table 2.1).

(ii) *Waste Disposal* (WD) of the chemical-related industrial and municipal waste. Waste can be generated either during the chemical lifecycle (e.g., from the production and industrial processes stages) or at the end of the lifespan of

Table 2.1 Descriptions and examples of lifecycle stages

	Description	Example
Production (LC1)	Manufacturing of a substance	Chemical synthesis, purification, drying, packaging, and transport from manufacturing sites, etc.
Industrial processes (LC2)	Mixing and blending a chemical with other chemicals to make a formulation, a preparation or a formulated product. Incorporating a chemical and/or its formulation into finished products or articles	Mixing multiple ingredients in personal care products; incorporating flame-retardants into plastics used in electrical and electronic equipment
Instantaneous use (LC3)	An *instantaneous* process in which a product or an article is applied, installed, assessed, or set up for an industrial or private (household) purpose	Synthetizing polymers using a processing aid, after which the processing aid is no longer needed; painting a wall using interior paints; applying personal care products or cosmetics
In service (LC4)	A *continuous* process that chemical-containing products or articles remain in use over the lifespan (usually longer than one year). In service stage is also referred to as the use phase	Long-life electrical and electronic equipment placed in the environment
Obsoleting and discarding (LC5)	Chemical-containing products or articles are disused, discarded, or dismantled	Demolition of buildings

commodities (i.e., from the in-service stage). CiP-CAFE considers five disposal approaches for *general* industrial and end-of-life waste: landfill (WD1), wastewater treatment (WD2), dumping and simple landfill (WD3), municipal solid waste incineration (WD4), inadvertent recycling of a CiP along with recycling or reuse of municipal solid waste (WD5). These five disposal approaches are generally applicable to all CiPs in all regions. CiP-CAFE additionally considers two disposal approaches for *CiP-specific* end-of-life waste: an inappropriate disposal approach that should be avoided (WD6, e.g., rudimentary dismantling and open burning of WEEE) and an environmentally sound disposal approach (WD7, e.g., best available technique or best environmental practice recommended by the Stockholm or Basel Conventions). These two disposal approaches are optional for certain CiPs and certain regions (Table 2.2).

(iii) *Environment* (E) that involves three receiving compartments of the chemical: atmosphere (E1), hydrosphere (E2) and pedosphere (E3). Other relevant receiving compartments, e.g., indoor air and human skin, can also be included in CiP-CAFE to describe the anthropospheric fate of chemicals used indoors.

If a stage or an environmental compartment is absent in a region, the model allows it to be parameterized as zero to remove it from the model calculations.

The totality of a chemical residing in a process is described as a *stock* (Fig. 2.1b) [16]. A process is *continuous* if the stock is not always zero, or otherwise, *instantaneous* (or *transient*) if the stock keeps zero or is depleted within the minimum time step in the calculation (a year in the case of CiP-CAFE). In this book, the accumulation of chemicals in in-service commodities (LC4) is termed an *in-use stock*. The accumulation of chemicals in waste disposal facilities, mainly landfill (WD1) and dumping and simple landfill (WD3), is termed a *waste stock*.

2.2.2 Calculation Modes

CiP-CAFE incorporates two calculation modes [26] to quantify flows and stocks of CiPs in the anthroposphere:

(i) A *top-down* mode starts the calculation from the Production (LC1) process. The most relevant input information is the annual production (or commercial

Table 2.2 Descriptions and examples of waste disposal approaches

	Description	Examples
Landfill (WD1)	Depositing industrial and municipal solid waste that contains a chemical into specially engineered landfill sites, which are "lined discrete cells which are capped and isolated from one another and the environment" (Annex I of EU Waste Framework Directive [18])	Landfilling of scrap plastics generated from industrial processes or waste plastic products
Wastewater treatment (WD2)	Treating industrial and municipal wastewater in wastewater treatment plants	Industrial and municipal wastewater treatment
Dumping and simple landfill (WD3)	Dumping industrial and municipal solid waste on open ground or deposit waste into poorly equipped landfill sites, which is usually unauthorized and hence illegal. For simplification, composting is also included in this class	"Fly-tipping" of construction and demolition waste in parks by local residents and construction and landscaping contractors

(continued)

Table 2.2 (continued)

	Description	Examples
Incineration (WD4)	Decomposing chemicals indiscriminately with general municipal solid waste in special furnaces with high temperature, which "involves heating to a minimum temperature of 850 °C with a residence time in the gas phase of at least two seconds" [19]. After combustion, cleaning equipment collects and destroys raw flue gases. Note that it differs from hazardous-waste incineration, e.g., incineration at a temperature greater than 1100 °C [19], which often destructs chemicals more completely and thus can be classified into WD7	Energy recovery from the combustion of municipal solid waste
Recycling and reuse (WD5)	Unintentional re-introduction of the chemical of interest back to the lifecycle phase during the recycling or reuse of recyclable non-hazardous materials in products or articles without complete elimination of the chemical of interest	Inadvertent introduction of brominated flame retardants to articles that are made from recycled plastics and do not need to be flame retarded [20–24]
Inappropriate treatment (WD6)	Illegal, unsafe, and uncontrolled disposal approaches that lead to severe contamination	Manual dismantling and open burning of waste electrical and electronic equipment [25]
Environmentally-sound management (ESM; WD7)	Managing chemical-containing waste in a manner causing zero or negligible emissions, i.e., "taking all practicable steps to ensure that hazardous wastes or other wastes are managed in a manner which will protect human health and the environment against the adverse effects which may result from such wastes" (the Basel Convention definition)	Methods and technologies for pre-treatment, destruction or irreversible transformation that are recommended as best available techniques (BATs) or best environmental practices (BEPs) [19]

tonnage) of a chemical in individual regions. It assumes that there are no chemical stockpiles awaiting sale in the Production stage (LC1) because most CiPs are manufactured in a build-to-order manner. This calculation mode relates the change of chemical mass within individual LC or WD processes with the flows that enter and depart from the process using first-order differential equations. For processes in the lifecycle phase, CiP-CAFE considers inflows including (i) mass flows along the product chain from the last process (for LCs 2–5) and (ii) recovered chemical via recycling and reuse (for LC2), and outflows including (i) mass flows along the product chain to the next process (for LCs 1–4), (ii) emissions, waste and decomposition (for LCs 1–5), (iii) obsolescence (for LC4). For processes in the waste disposal phase, CiP-CAFE considers inflows including (i) waste flows from the lifecycle phase (for WDs 1–7) and (ii) obsolescence (for WDs 1–7), and outflows including (i) emissions, waste and dissipation (e.g., decomposition and degradation) (for WDs 1–7), and (ii) recycling and reuse (for WD5). For mathematical expressions of individual processes, readers are recommended to refer to Li and Wania [27].

(ii) A *bottom-up* mode starts the calculation from the in-service (LC4) process. This calculation mode first calculates the in-use stocks by summing up the chemical contents in commodities in five applications (Eq. 2.2):

$$M(t)_{RE(i),LC4} = \sum_{k=1}^{5} P(t)_{RE(i),AP(k)} \cdot C(t)_{RE(i),AP(k)} \tag{2.2}$$

where, $P(t)_{RE(i),AP(k)}$ and $C(t)_{RE(i),AP(k)}$ are the number of commodities (e.g., personal computers) and the chemical contents therein (e.g., PCB concentrations in the plastics of a personal computer) in application k in region i, respectively. Based on the estimated in-use stocks, CiP-CAFE then calculates forward the masses in WD processes using the same method as the top-down mode and traces back the masses in LC1 through LC3.

It is challenging to obtain analytical solutions for differential equations in CiP-CAFE. Therefore, it is solved numerically by a finite difference scheme. The model adopts a time step of one year because statistics on most merchandise production and trade are published on an annual basis.

2.2.3 Configuration and Parameterization

This section gives an overview of the mathematical expressions of individual inflows (N_{in}) and outflows (N_{out}), including net import flows (Sect. 2.2.3.1), emission, waste and decomposition (Sect. 2.2.3.2), and obsolescence from in-use stock to waste stream (Sect. 2.2.3.3). We also briefly indicate possible sources for retrieving inputs parameters.

2.2.3.1 Transboundary Transport of Chemicals via Interregional Trade

A chemical can be subject to transboundary transport between regions as a technical substance after production (LC1), within commodities after industrial processes (LC2) and/or within waste at the end of the service life (LC5). For a lifecycle stage j ($j = 1, 2,$ and 5), the *net* amount of a chemical entering the region i ($i = 1, 2, ...,$ 7) via interregional trade, $INT(t)$, is expressed as the difference between the import $IMP(t)$ and export $EXP(t)$ in each application k ($k = 1, 2, ..., 5$) (Eq. 2.3).

$$INT(t)_{RE(i),LC(j),AP(k)} = IMP(t)_{RE(i),LC(j),AP(k)} - EXP(t)_{RE(i),LC(j),AP(k)} \qquad (2.3)$$

A region imports a chemical if $INT(t)$ is calculated to be positive and otherwise exports a chemical if $INT(t)$ is calculated to be negative. In particular, the $INT(t)$ of technical substance after LC1 can be viewed as the difference between regional *production* and *consumption* (strictly speaking, the "apparent consumption") of a chemical. The export and import are calculated using Eqs. (2.4) and (2.5).

$$EXP(t)_{RE(i),LC(j),AP(k)} = \begin{cases} M(t)_{RE(i),LC(j),AP(k)} \cdot EXPF(t)_{RE(i),LC(j),AP(k)}, j = 1, 2 \\ DIS(t)_{RE(i),AP(k)} \cdot EXPF(t)_{RE(i),LC5,AP(k)}, j = 5 \end{cases} \qquad (2.4)$$

$$IMP(t)_{RE(i),LC(j),AP(k)} = \sum_{i=1}^{7} EXP(t)_{RE(i),LC(j),AP(k)} \cdot IMPF(t)_{RE(i),LC(j),AP(k)} \qquad (2.5)$$

whereby $M(t)$ denotes chemical mass in a lifecycle stage; $DIS(t)$ denotes chemical mass entering the waste stream from in-use stocks (see Sect. 2.2.3.3). $EXPF(t)$ refers to the fraction of technical substance, commodities or waste exported from a region, and $IMPF(t)$ refers to the fraction of technical substance, commodities and waste imported to region i in the total international trade, i.e., the sum of regional exports.

Underlying Eqs. (2.4) and (2.5) is an assumption that the fractions of a chemical exported with commodities and waste are equal to the fractions of commodities and waste exported. In other words, chemical-containing commodities and/or waste are indistinguishable from their chemical-free counterparts in a perfectly competitive market. This assumption is justifiable because, during most of the commercial history of a CiP, chemical-containing commodities or waste are usually traded without visual labeling or notification; international trade restrictions are enforced only after its adverse environmental effects have been disclosed and practical fast detection technique becomes available. The $EXPF(t)$ and $IMPF(t)$ of technical substances are often available from manufacturers and distributors, and those of finished products and waste are retrievable from international trade statistics (e.g., the World Trade Organization International Trade Statistics, UN Comtrade database, and UN Industrial Commodity Statistics Database) and/or peer-reviewed literature (e.g., international trade of e-waste reported by Breivik et al. [28]).

2.2.3.2   Emission, Waste and Decomposition

In CiP-CAFE, the emission of a chemical to an environmental compartment p in region i, $EMI(t)$, is proportional to the activity level in the lifecycle stage j, $ACT(t)$. The coefficient of the proportionality is defined as an emission factor $EF(t)$ (Eq. 2.6)

$$EMI(t)_{RE(i),LC(j),AP(k),ENV(p)} = ACT(t)_{RE(i),LC(j),AP(k)} \times EF(t)_{RE(i),LC(j),AP(k),ENV(p)}.$$
$$(2.6)$$

Likewise, we can calculate the emission from a waste disposal activity m using Eq. (2.7):

$$EMI(t)_{RE(i),WD(m),ENV(p)} = ACT(t)_{RE(i),WD(m)} \times EF(t)_{RE(i),WD(m),ENV(p)}. \qquad (2.7)$$

In this way, we can also define waste factors $WF(t)$ that quantify chemical mass entering the waste stream, and decomposition factors $DF(t)$ that quantify chemical mass decomposed or degraded.

In a *continuous* process, the activity level $ACT(t)$ equates to the stock within the process and these factors have units of tonne-emissions per tonne-stock per annum; while in an *instantaneous* process, the activity level $ACT(t)$ equates to the amount of a chemical that enters the process and these factors have units of tonne-emissions per tonne-inflow.

Emission, waste and decomposition factors for transient processes are usually well characterized, compiled and tabulated in the literature or in databases. Typical examples include the EU Technical Guidance Document on Risk Assessment [17], the OECD Emission Scenario Documents [29], and the EU Specific Environmental Release Categories (SPERCs) [30]. In contrast, the factors for continuous processes are often unavailable. In order to facilitate estimating emission factors for continuous sources based on regional environmental characteristics (e.g., average annual temperature and precipitation), CiP-CAFE incorporates three optional, built-in modules:

(i)   the EmissionRate module simulates emissions from in-service articles placed in the environment to the air above the product surface (i.e., for LC4) [27];

(ii)   the Model for Organic Chemicals in Landfills (MOCLA) [31] module simulates emissions from landfill (WD1) and dumping and simple landfill (WD3) to air, freshwater and soil, and degradation in WD1 and WD3. For WD1, given that an engineered landfill is often equipped with a pre-treatment system, an impermeable bottom liner, a leachate drainage, and landfill gas collection system, CiP-CAFE assumes that 50% of generated landfill gas and 80% of generated leachate will be collected, completely treated and not emitted into the environment, in accordance with the assumptions in Nielsen and Hauschild [32]. For WD3, CiP-CAFE assumes that all generated landfill gas and leachate enter the environment; and

(iii) the SimpleTreat module [33] simulates emissions from wastewater treatment plants (WD2) to air, freshwater and soil, and degradation in wastewater treatment plants. CiP-CAFE assumes that all chemical evaporated from wastewater treatment plants enters the air compartment, chemical in effluent being 100% discharged into the receiving freshwater, and 50% of sewage sludge is returned to the soil.

When indoor emissions and skin application are considered in CiP-CAFE, if desired, users can also use external models, e.g., a product intake fraction model [34], to calculate emission factors to these two compartments.

For each chemical, partition coefficients K_{OW}, K_{OA}, and K_{AW} and corresponding internal energies for phase transfer (U_{OW}, U_{OA}, and U_{AW}), and degradation half-lives in the air (HL_A), solid (HL_S) and water (HL_W) phases and corresponding activation energies for degradation are required as inputs to drive these modules. Region-specific average annual temperature and precipitation are contained in CiP-CAFE as built-in data.

2.2.3.3 Obsolescence from in-Use Stock to the Waste Stream

For each application, the chemical mass entering the waste stream along with obsolete products, $DIS(t)$, is calculated based on the lifespan distribution function of *the chemical* $f_{chemical}(i)$ and the annual production (similar to Eq. 1.3 in Chap. 1.1). In CiP-CAFE, $f_{chemical}(i)$ is calculated from the lifespan distribution function of *the product* $f_{product}(i)$ (Eq. 1.2 in Chap. 1.1) with considering the cumulative loss through emission, waste and decomposition of the chemical during the product lifespan (i.e., process LC4):

$$f_{chemical}(i) = f_{product}(i) \cdot \prod_{\tau=1}^{i} \left[1 - \sum_{p} EF_p(\tau) - \sum_{q} WF_q(\tau) - DF(\tau) \right], \quad (2.8)$$

where $\sum_{i=1}^{+\infty} f_{prod}(i) = 1$ by its definition. An inspection of Eq. (2.8) demonstrates that the lifespan of a chemical is not necessarily equal to that of corresponding products, in particular in the case that the chemical loss is pronounced during the product lifespan. Equation (2.8) reflects the difference between anthropospheric fate modeling for chemicals and products.

There are four types of built-in lifespan distribution functions in CiP-CAFE (Table 2.3), which have been well-defined and utilized in dozens of previous studies [26, 35]. A commonly required input for the four lifespan distribution functions is a "central parameter" that describes the central tendency of a lifespan (Table 2.3), which is a statistical measure of the expectancy, or the average, of a lifespan. Readers are recommended to pay particular attention to the definition of the average lifetime; for instance, a normally distributed lifetime with an average lifespan of 10 years means that 50% of the products, instead of all, are cumulatively discarded by the

Table 2.3 Lifespan distribution functions defined in the CiP-CAFE

Distribution	Probability density function	Central parameter	Deviation parameter
Normal	$f(t) = \frac{1}{\sigma\sqrt{2\pi}} \exp\left[-\frac{1}{2}\left(\frac{t-\mu}{\sigma}\right)^2\right]$	μ–average lifespan	σ–standard deviation
Weibull	$f(t) =$ $\frac{k}{T} \cdot \left(\frac{t-0.5}{T}\right)^{k-1} \cdot \exp\left(-\frac{t-0.5}{T}\right)^k$	T–scale parameter	k–shape parameter
Delta	$f(t) = \begin{cases} 0, & t \neq t_m \\ 1, & t \neq t_m \end{cases}$	t_m–fixed lifespan	
Bathtub	$f(t) =$ $\begin{cases} a_0(3e^{-t} + 1), & t \geq T \\ a_0(3e^{t-2T} + 1), & T < t \leq 2T \end{cases}$ where a_0 satisfies $\int_0^{2T} f(t) \cdot dt = 1$	T–use-life expectancy	

end of the 10th year. Some lifespan distribution functions additionally require a "deviation parameter" to describe how variable a lifespan can be.

While CiP-CAFE allows a single $f_{chemical}(i)$ for each application, an application actually comprises a number of products with different $f_{chemical}(i)$ because different product models, product classes and product forms possess diverse product lifespans and emission, waste and decomposition behavior. Since most lifespan distributions are additive, i.e., the sum of multiple lifespans following a certain statistical distribution still follows the same distribution, it is always mathematically possible to find an "overall" lifespan distribution to aggregate the lifespan distributions of multiple products.

2.2.3.4 Distribution of a Chemical Among Waste Disposal Approaches

At the end of their lifespan, commodities are discarded and disposed of, both in the same way as general waste (WDs 1, 3, 4 and 5) and in a CiP-specific manner (WDs 6 and 7). CiP-CAFE uses a "waste disposal ratio" to characterize the fraction of total waste distributed to each waste disposal approach. The model contains statistical data on the time-dependent (from 1930 to 2100), region-specific waste disposal ratios for general solid waste disposed of in WDs 1, 3, 4 and 5 [27], and assumes all general liquid waste to enter WD 2. For each region, when applicable, the user needs to input application-specific waste disposal ratios for two CiP-specific waste disposal approaches, i.e., WD6 and WD7. The model then allocates the rest of waste to WDs 1, 3, 4 and 5 based on the built-in waste disposal ratios if it is solid, or to WD 2 if it is liquid.

2.2.4   Model Outputs

The CiP-CAFE outputs comprise annual estimates of stocks, flows and emissions of the compound of interest in individual applications and regions.

In order to be compatible with environmental fate models, CiP-CAFE spatially extrapolates the emission estimates by region and process into a $1° \times 1°$ longitude/latitude grid system using the methodology introduced by Li et al. [36] and default surrogate data of

(i)   user-input geographically explicit information on annual production (or production capacity) for emissions from production (LC1),
(ii) geographic distribution of the gross domestic product (GDP) for emissions from industrial processes (LC2) and from the disposal of industrial waste (WDs 1, 2 and 3),
(iii) geographic distribution of population density for emissions from instantaneous use (LC3), in service (LC4) and obsoleting and discarding (LC5), and from the disposal of waste arising from LCs 3, 4 and 5 (WDs 1, 2, 3, 4, 5 and 7),
(iv) geographic distribution of the reciprocal of GDP per capita for emissions from inappropriate treatment, based on the assumption that the less developed a region is, the more likely that residents handle the waste in a primitive manner.

The geographically explicit GDP and population data are averages between 1990 and 2025 of the B2 scenarios defined according to the Special Report on Emissions Scenarios (SRES) by the Intergovernmental Panel on Climate Change (IPCC).

2.3   BETR-Global Model: Environmental Fate Modeling

In this section, we will give a brief overview of the BETR-Global model, which is used for environmental fate modeling. The development of the BETR-Global model was a joint effort between the Lawrence Berkeley National Laboratory (BE) in the US and Trent University (TR) in Canada. For details, the readers are recommended to refer to MacLeod et al. [37, 38].

2.3.1   Structure and Conventions

The BETR-Global model describes the global physical environment as 288 multimedia regions on a $15° \times 15°$ longitude/latitude grid (Fig. 2.2). Each grid cell is divided into seven interconnected bulk compartments: (i) upper air, (ii) lower air, (iii) vegetation, (iv) freshwater, (v) coastal water, (vi) soil and (vii) freshwater sediment. Such a seven-compartment segmentation is a typical characteristic of BETR model series such as BETR North America [39] and BETR-World [40]. If a compartment

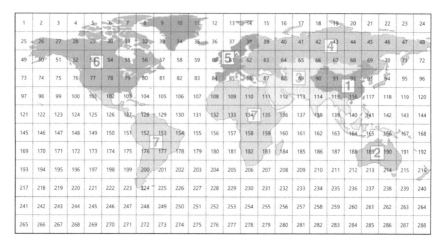

Fig. 2.2 The grid in BETR-global and the relationship with the region numbers in CiP-CAFE

is absent in a grid cell, the model allows its volume to be set to zero to remove it from the model calculation.

A modeled chemical diffuses between the bulk compartments, reacts within these bulk compartments, and migrates between grid cells along with the advection of air and water compartments. For reaction, the model requires the user to input degradation half-lives in individual bulk compartments and corresponding activation energies for degradation. For advection, the model considers the mass balance of water and air between grid cells.

Each bulk compartment is further segmented into several immiscible phases such as the gaseous, water, organic matter and solid phases. The partition of chemicals between the phases are expressed as functions of partition coefficients K_{OW}, K_{OA}, and K_{AW}. Internal energies for phase transfer, U_{OW}, U_{OA}, and U_{AW}, are required for temperature adjustment.

BETR-Global supports Level III and Level IV simulations. The Level IV mode will be employed for calculations in this book. The structure of the physical environment, e.g., wind speed, temperature, OH radical concentration, precipitation rate, and atmospheric and oceanic circulations in individual grid cells, represents the average status between 1960 and 1999. These structural data are georeferenced reanalyzed data from the literature; for details, please consult the BETR-Global website.[1]

The outputs of BETR-Global include concentrations and masses (termed "inventory" in the model) of chemicals in individual compartments of individual cells, as well as the mass exchanges between compartments and between cells. The user can customize the time step for outputs, which is set as three months in all calculations performed in this book. Three months is believed to be appropriate because it agrees with the sampling duration of most passive samplers, which provide measured concentration data for evaluating our model performance.

---

[1] Please visit https://sites.google.com/site/betrglobal/home/model-parameterization.

2.3.2 Emissions: A Bridge Between CiP-CAFE and BETR-Global

BETR-Global is fed with emission estimates from CiP-CAFE. The annual emissions to the atmosphere, hydrosphere and pedosphere generated by CiP-CAFE are input to the lower air, freshwater and soil, respectively, in BETR-Global. The CiP-CAFE-derived $1° \times 1°$ gridded emission estimates are aggregated to the $15° \times 15°$ resolution used by BETR-Global. For lower air emissions, BETR-Global redistributes the annual emissions to a monthly resolution based on the average temperature of individual months $(T_1, T_2, …, T_{12})$ and internal energy $(U_{OA}$; Eq. 2.9); for emissions to other environmental compartments, BETR-Global redistributes the annual numbers equally to individual months [37].

$$E_{air,i} = E_{air,annual} \cdot \frac{\exp\left(\frac{U_{OA}}{R \cdot T_i}\right)}{\exp\left(\frac{U_{OA}}{R \cdot T_1}\right) + \exp\left(\frac{U_{OA}}{R \cdot T_2}\right) + \cdots + \exp\left(\frac{U_{OA}}{R \cdot T_{12}}\right)} \tag{2.9}$$

where $E_{air,i}$ denotes the calculated monthly emission in month i, $E_{air,annual}$ represents the annual air emission, and R is the gas constant.

2.4 Summary

In this chapter, we describe the rationale, structure and configuration of the developed anthropospheric fate model, CiP-CAFE. CiP-CAFE is coupled with an environmental fate model named BETR-Global, based on an analogy between chemical anthropospheric and environmental fate modeling. This modeling framework allows simulations from chemical production to concentrations in environmental compartments. In the following chapters, we will illustrate the powerful applications of this modeling framework in addressing the issue of chemicals in products.

References

1. Mackay D (2001) Multimedia environmental models: the fugacity approach, 2nd edn. CRC Press, Boca Raton, FL
2. Ayres RU, Ayres LW (1998) Accounting for resources 1: economy-wide applications of mass-balance principles to materials and waste. Edward Elgar, Cheltenham, UK and Lyme, MA
3. Ayres RU, Kneese AV (1969) Production, consumption, and externalities. Am Econ Rev 59(3):282–297
4. Breivik K, Armitage JM, Wania F, Sweetman AJ, Jones KC (2016) Tracking the global distribution of persistent organic pollutants accounting for e-waste exports to developing regions. Environ Sci Technol 50(2):798–805

5. Wania F, Daly GL (2002) Estimating the contribution of degradation in air and deposition to the deep sea to the global loss of PCBs. Atmos Environ 36(36–37):5581–5593

6. Schellenberger S, Gillgard P, Stare A, Hanning A, Levenstam O, Roos S, Cousins IT (2018) Facing the rain after the phase out: performance evaluation of alternative fluorinated and non-fluorinated durable water repellents for outdoor fabrics. Chemosphere 193:675–684

7. Mackay D, Celsie AK, Parnis JM (2015) The evolution and future of environmental partition coefficients. Environ Rev 24(1):101–113

8. Wania F (2003) Assessing the potential of persistent organic chemicals for long-range transport and accumulation in polar regions. Environ Sci Technol 37(7):1344–1351

9. Wania F (2006) Potential of degradable organic chemicals for absolute and relative enrichment in the Arctic. Environ Sci Technol 40(2):569–577

10. Schwarzenbach RP, Gschwend PM, Imboden DM (2003) Environmental organic chemistry, 2nd edn. Wiley & Sons Inc, Hoboken, NJ

11. Li L, Wania F (2018) Occurrence of single- and double-peaked emission profiles of synthetic chemicals. Environ Sci Technol 52(8):4684–4693

12. Murakami S, Oguchi M, Tasaki T, Daigo I, Hashimoto S (2010) Lifespan of commodities, part I: the creation of a database and its review. J Ind Ecol 14(4):598–612

13. Sigman R, Hilderink H, Delrue N, Braathen NA, Leflaive X (2012) Health and environment. In: Organisation for Economic Co-operation and Development (OECD) (ed) OECD environmental outlook to 2050: the consequences of inaction. OECD Publishing, Paris, France, pp 275–332

14. UNEP (2019) Global chemicals outlook II. From legacies to innovative solutions: implementing the 2030 agenda for sustainable development. United Nations Environment Programme, Nairobi

15. Baccini P, Brunner PH (2012) Metabolism of the anthroposphere: analysis, evaluation, design, 2nd edn. The MIT Press, Cambridge, MA and London

16. Pauliuk S, Majeau-Bettez G, Müller DB, Hertwich EG (2015) Toward a practical ontology for socioeconomic metabolism. J Ind Ecol 20(6):1260–1272

17. ECB (2003) Technical guidance document on risk assessment. European Chemicals Bureau (ECB), Italy

18. EU (2008) Directive 2008/98/EC of the European parliament and of the council of 19 November 2008 on waste and repealing certain directives

19. Secretariat of the basel convention (2017) General technical guidelines for the environmentally sound management of wastes consisting of containing or contaminated with persistent organic pollutants. Secretariat of the Basel Convention, Geneva

20. Puype F, Samsonek J, Knoop J, Egelkraut-Holtus M, Ortlieb M (2015) Evidence of waste electrical and electronic equipment (WEEE) relevant substances in polymeric food-contact articles sold on the European market. Food Addit Contam Part A 32(3):410–426

21. Samsonek J, Puype F (2013) Occurrence of brominated flame retardants in black thermo cups and selected kitchen utensils purchased on the European market. Food Addit Contam Part A 30(11):1976–1986

22. Abdallah MA-E, Sharkey M, Berresheim H, Harrad S (2018) Hexabromocyclododecane in polystyrene packaging: a downside of recycling? Chemosphere 199:612–616

23. Rani M, Shim WJ, Han GM, Jang M, Song YK, Hong SH (2014) Hexabromocyclododecane in polystyrene based consumer products: an evidence of unregulated use. Chemosphere 110:111–119

24. Kuang J, Abdallah MA-E, Harrad S (2018) Brominated flame retardants in black plastic kitchen utensils: concentrations and human exposure implications. Sci Total Environ 610:1138–1146

25. Zhang K, Schnoor JL, Zeng EY (2012) E-waste recycling: where does it go from here? Environ Sci Technol 46(20):10861–10867

26. Müller E, Hilty LM, Widmer R, Schluep M, Faulstich M (2014) Modeling metal stocks and flows: a review of dynamic material flow analysis methods. Environ Sci Technol 48(4):2102–2113

27. Li L, Wania F (2016) Tracking chemicals in products around the world: introduction of a dynamic substance flow analysis model and application to PCBs. Environ Int 94:674–686

28. Breivik K, Armitage JM, Wania F, Jones KC (2014) Tracking the global generation and exports of e-waste. Do existing estimates add up? Environ Sci Technol 48(15):8735–8743
29. OECD (2000) Guidance document on emission scenario documents (ENV/JM/MONO(2000) 12). OECD series on emission scenario documents. Environment Directorate, Organisation for Economic Co-operation and Development, Paris
30. Sättler D, Schnöder F, Aust N, Ahrens A, Bögi C, Traas T, Tolls J (2012) Specific environmental release categories—a tool for improving chemical safety assessment in the EC–Report of a multi-stakeholder workshop. Integr Environ Assess Manag 8(4):580–585
31. Kjeldsen P, Christensen TH (2001) A simple model for the distribution and fate of organic chemicals in a landfill: MOCLA. Waste Manag Res 19(3):201–216
32. Nielsen PH, Hauschild M (1998) Product specific emissions from municipal solid waste landfills. Part I: Landfill model. Int J Life Cycle Assess 3(3):158–168
33. Struijs J (2014) SimpleTreat 4.0: a model to predict fate and emission of chemicals in wastewater treatment plants: background report describing the equations (RIVM Rapport 601353005). National Institute for Public Health and the Environment (RIVM), Netherland, Bilthoven
34. Csiszar SA, Ernstoff AS, Fantke P, Jolliet O (2017) Stochastic modeling of near-field exposure to parabens in personal care products. J Expo Sci Environ Epidemiol 27(2):152
35. Oguchi M, Murakami S, Tasaki T, Daigo I, Hashimoto S (2010) Lifespan of commodities, Part II: methodologies for estimating lifespan distribution of commodities. J Ind Ecol 14(4):613–626
36. Li Y-F, McMillan A, Scholtz MT (1996) Global HCH usage with 1×1 longitude/latitude resolution. Environ Sci Technol 30(12):3525–3533
37. MacLeod M, Riley WJ, Mckone TE (2005) Assessing the influence of climate variability on atmospheric concentrations of polychlorinated biphenyls using a global-scale mass balance model (BETR-Global). Environ Sci Technol 39(17):6749–6756
38. MacLeod M, von Waldow H, Tay P, Armitage JM, Wöhrnschimmel H, Riley WJ, McKone TE, Hungerbuhler K (2011) BETR global—a geographically-explicit global-scale multimedia contaminant fate model. Environ Pollut 159(5):1442–1445
39. MacLeod M, Woodfine DG, Mackay D, McKone T, Bennett D, Maddalena R (2001) BETR North America: a regionally segmented multimedia contaminant fate model for North America. Environ Sci Pollut Res 8(3):156–163
40. Toose L, Woodfine DG, MacLeod M, Mackay D, Gouin J (2004) BETR-World: a geographically explicit model of chemical fate: application to transport of α-HCH to the Arctic. Environ Pollut 128(1–2):223–240

Part II
Case Studies

Chapter 3
Global Long-Term Fate and Dispersal of Polychlorinated Biphenyls

Abstract In this chapter, we explore (i) the temporal trend of stocks and emissions of polychlorinated biphenyls (PCBs) throughout the lifecycle of PCB-containing products, and (ii) the spatial dispersal of PCBs through international trade of technical mixtures, finished products and waste (anthropospheric dispersal), as well as long-range environmental transport mediated by air and ocean currents (environmental dispersal). We reveal that in-use and waste stocks are temporary sinks for PCBs in the anthroposphere and relevant sources of PCBs to the global physical environment. After a sufficiently long period of time, PCBs either permanently disappear from lifecycle and waste disposal phases or enter the physical environment. In addition, the anthropospheric dispersal is more efficient than the environmental dispersal in driving the global dispersal of PCBs. The role of the international trade of technical mixtures is more prominent than that of other trade flows.

3.1 Introduction

There seems to be no chemical that is more lingering and ubiquitous than polychlorinated biphenyls (PCBs), which are a class of synthetic chemicals widely used in electrical and electronic equipment, sealants, rubbers and other materials because of perfect electrical and thermal isolation performance and high chemical stability. While the manufacturing and use of these compounds had ceased in most countries around the world before the 2000s [1], fresh emissions of PCBs are still ongoing and even considerable in some regions. In the meantime, while the production of these compounds had been restricted within no more than 15 countries [1], PCBs were detected in environmental media or biota in almost every corner on our planet, even in remote background regions such as the Arctic and Antarctic [2–5].

Emissions of PCBs to the environment occur throughout the lifespans of PCB-containing products. Earlier monitoring evidence has demonstrated that industrial processes are a relevant source responsible for the extreme contamination observed in hotspot regions [6, 7]. In a few countries, emissions from the landfills or dumps of legacy industrial waste (i.e., waste stocks) still contribute to the environmental occurrence of PCBs, even decades after the shutdown of PCB production and processing

© Springer Nature Singapore Pte Ltd. 2020
L. Li, *Modeling the Fate of Chemicals in Products*, Springer Theses,
https://doi.org/10.1007/978-981-15-0579-9_3

facilities [8]. On the other hand, emissions of PCBs from products or materials still in service (i.e., in-use stocks) have been confirmed as another important source to their presence in the environment [9, 10]. Considerable amounts of PCBs also originate from the end-of-life disposal of PCB-containing waste. As such, to minimize or curb the long-term emissions of PCBs, it is essential to elucidate the relative importance of emissions from different lifecycle stages of PCB-containing products, and in particular, how the relative importance evolves with time in a scale of decades or centuries.

PCBs can disperse globally through multiple vectors in the anthroposphere and environment. In the anthroposphere, PCBs can enter a country via importation of the substance (i.e., the technical PCB mixture), PCB-containing finished products (e.g., transformers and capacitors), as well as PCB-containing waste (e.g., waste electrical and electronic equipment). Such trade-mediated long-range transport, in particular, illegal trade and primitive disposal of PCB-containing waste, explains the observed PCB occurrence in the West African region, where production and consumption are rather rare [11]. On the other hand, PCBs can also undergo long-range transport mediated by air and ocean currents. For example, warm temperatures favor evaporation of PCBs from surface environmental media in tropical or subtropical regions, whereas cool temperatures facilitate deposition of PCBs from the atmosphere and enrichment in soil and water bodies once PCBs reach high latitude regions [12]. In addition, migratory organisms, such as birds, fish and marine mammals, can also deliver PCBs over long distances and across international boundaries, despite minimal efficiency compared with the abiotic vectors [13]. Therefore, it would be interesting to establish whether the anthropospheric or environmental vector is more relevant for the worldwide dispersal of PCBs.

In this chapter, we will address the two questions above using combined anthropospheric (CiP-CAFE) and environmental (BETR-Global) fate modeling. Specifically, we will first provide a comprehensive overview of the major pathways that PCBs are taking through the anthroposphere. We then analyze (i) the size and temporal evolution of in-use and waste stocks, and emissions arising therefrom, and (ii) the extent to which international trade of substance, products, and waste, contributes to the global dispersal of PCBs, and its magnitude compared to global-scale environmental transport.

3.2 Methods and Data

3.2.1 Properties of PCB Congeners

PCBs comprise 209 congeners with different numbers of chlorine substituents and different positions of chlorine atoms. In this chapter, we simulate the anthropospheric and environment fates of six most frequently monitored and investigated congeners (collectively referred to as Σ_6PCBs hereafter): diCB-8 (CAS no. 34883-43-7), triCB-28 (CAS no. 7012-37-5), tetraCB-52 (CAS no. 35693-99-3), pentaCB-118 (CAS no. 31508-00-6), hexaCB-153 (CAS no. 35065-27-1) and heptaCB-180 (CAS no. 35065-29-3).

For each congener, the partition coefficients (K_{AW}, K_{OW} and K_{OA}) at 25 °C are recommended values from Li et al. [14], which have been adjusted for thermodynamic consistency. Internal energies of phase transfer (ΔU_W, ΔU_A and ΔU_O), which are used to adjust the partition coefficients for temperature dependence, are calculated in accordance with MacLeod et al. [15] based on Trouton's Rule and defaults. Degradation half-lives in air, water (and wastewater), soil, and sediment (and solid waste in landfill, and dumping and simple landfill) at 7 °C are taken from Sinkkonen and Paasivirta [16]. Activation energies for adjusting the degradation half-lives for temperature dependence are defaults in the CiP-CAFE and BETR-Global models.

3.2.2 Production, Applications and Interregional Trade

Worldwide PCB production lasts from 1930 to 1993 [1]. In this chapter, the annual productions of the six congeners in regions except for mainland China (Region 1) are taken from Breivik et al. [1], whereas the annual production of each congener in mainland China (Region 1) is calculated as a product of annual production of the technical PCB mixtures in China and the respective mass fraction in the technical PCB mixtures [17]. The compiled data show that cumulatively 303 kilotonnes (kt) of Σ_6PCBs have been produced between 1930 and 1993, with PCB-8 (26%), PCB-28 (23%) and PCB-118 (16%) as three most abundant congeners, and Western Europe (42%) and North America (40%) as two main producers. It should be noted that, in addition to intentional synthesis, PCBs can be unintentionally formed and released from industrial processes such as waste incineration and metal production. Given that the unintentionally generated PCBs account for less than 1% of the overall historical PCB emissions [18] and concentrate mainly within industrial areas [19], it is acceptable to omit the unintentionally generated PCBs in our global modeling.

We consider four main applications (APs) of PCBs here following Breivik et al. [20]: open usage (AP1), small capacitor usage (AP2), nominally closed usage (AP3), and closed transformer usage (AP4). The relative distribution of the consumption of the technical PCB mixture among the four applications in China is taken from Cui et al. [17], and the relative distributions in other regions are taken from Breivik et al. [20].

Breivik et al. [20] assumed an average lifespan of 5 years for products in AP1, 13 years for products in AP2, 10 years for products in AP3, and 25 years for products in AP4. However, since then new information on the behavior of PCBs in the anthroposphere has become available, most notably indicating that PCB-containing products might emit PCBs for much longer than previously estimated. For instance, in developing countries where environmentally sound treatment technologies are often not accessible or affordable, transformers containing PCBs (AP4) have to be temporarily stored outdoors after their decommission [21]. The temporary storage after decommission can be for up to another two decades [22], during which PCBs continue to be emitted at rates comparable to those occurring in the use phase. Meanwhile, in developed countries such as Canada [23] and Switzerland [24, 25], emissions from PCB-containing joint sealants (AP1) are still ongoing. Taking into account such information, we assume that (i) the product lifespan in AP1 is 25 years in Western Europe (Region 5) and North America (Region 6) but remains the same as Breivik et al. [20] in other regions; and (ii) that the product lifespan in AP4 is 45 years in mainland China (Region 1), Central and South Asia (Region 3), Central and Eastern Europe (Region 4) and the rest of the world (Region 7) but remains the same as Breivik et al. [20] in other regions. We assume that the product lifespans follow the bathtub distribution, following Breivik et al. [20].

We consider the interregional mass exchange of PCBs through trade flows of (i) the technical PCB mixture (after LC1; data taken from Breivik et al. [26]), (ii) PCB-containing finished products (after LC2; data taken from statistical databases [27–30]), and (iii) PCB-containing waste electrical and electronic equipment (WEEE) (after LC5; data taken from Breivik et al. [31]). The import and export fractions of (ii) and (iii) are assumed to be constant throughout the modeled period, due to a lack of temporally resolved data.

3.2.3 Emission, Waste and Decomposition Factors

Emission and waste factors for continuous processes are compiled from the literature, whereas those for instantaneous processes, i.e., in-service (LC4), landfill (WD1), wastewater treatment (WD2), and dumping and simple landfill (WD3), are computed using built-in modules in CiP-CAFE (Sect. 2.2.3.2). For details, see Li and Wania [32]. We do not consider the decomposition of PCB congeners during lifecycle stages.

3.2.4 Disposal of PCB-Containing Solid Waste

In accordance with Breivik et al. [33], we assume that 20% of the waste electrical and electronic products imported to developing regions (Regions 1, 3, and 7) is subject to illegal open burning, i.e., the inappropriate treatment (WD6) labeled in CiP-CAFE, throughout the simulation period. We further assume absence of environmentally sound treatment of PCB-containing waste (WD7). The relative distribution of the rest of waste among WDs 1, 3, 4 and 5 is based on the built-in time-dependent waste disposal ratios for individual regions in CiP-CAFE.

3.3 Evaluation of Modeling Performance

Prior to presenting and interpreting modeling results, we evaluate the performance of the combined anthropospheric and environmental fate modeling. We model air concentrations of the six congeners and compare the model predictions with measured concentrations collected from the literature. Here, air concentrations are chosen for comparison because they tend to mimic atmospheric emissions closely. In the following, the model predictions and measurements will be compared in terms of absolute levels (static snapshots) and long-term temporal trends (dynamic evolution).

First, we compare measured and modeled congener-specific concentrations for Western Europe (Cell 61 of the BETR-Global grid, Fig. 2.2), a region comprising urban areas that are heavily influenced by human activities as well as background areas that are barely affected. Figure 3.1 shows that (i) the measurement data points are evenly distributed on either side of the diagonal line (this line corresponds to perfect agreement between literature-reported and modeled concentrations), with urban areas falling above and remote areas falling below, and (ii) the difference between modeled and measured concentrations is mostly within an order of magnitude. The comparison shows that our modeled concentrations are representative of the regional average, although the measured concentrations vary substantially because of the geographic heterogeneity in the cell.

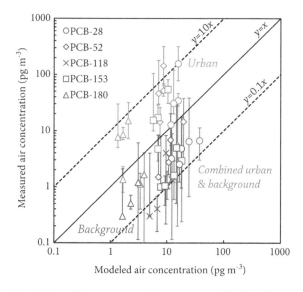

Fig. 3.1 Comparison between literature-reported measurements [24, 38–42] and modeled air concentrations for Western Europe (Cell 61 of the BETR-Global grid). Different colors of data points represent different categories of sampling sites. Monitoring campaigns encompassing both urban and remote sampling sites are denoted as "combined urban and background". Diagonal lines represent perfect agreement (solid) and agreement within an order of magnitude (dashed). Reproduced with permission (Li and Wania [32]) Copyright (2016) Elsevier

Table 3.1 Comparison between modeled (BETR-Global outputs for the period 1996–2010) and observed $t_{1/2,conc}$ for Cells 8, 13, 38 and 79 in the BETR-Global grid

Cell no.	Cell 8		Cell 13		Cell 38		Cell 79	
	Modeled	Observed[1]	Modeled	Observed[2]	Modeled	Observed[3]	Modeled	Observed[4]
ΣPCBs		N.A.		N.A.		N.A.		7.7–15
PCB28	8.1–11	7.1	7.9–10	8.2	8.0–10	15	6.3–9.5	N.A.
PCB52	8.1–13	4.6–9.8	7.8–13	4.4	7.9–13	17	6.1–12	N.A.
PCB118	10–18	8.3–23	8.8–18	6.1	8.7–18	11	6.4–19	N.A.
PCB153	15–36	8.3–16	13–33	6.2	11–28	11	8.3–27	N.A.
PCB180	9.1–30	3.6–17	8.2–29	4.4	8.0–28	5.9	5.6–25	N.A.

Notes
1. Data from Alert, Canada (1993–2012) [37];
2. Data from Zeppelin, Norway (1998–2012) [37];
3. Data from Pallas, Finland (1996–2012) [37];
4. Data from Chicago, Illinois (1997–2003) [35], Sleeping Bear Dunes, Michigan (1992–2001) [34], Sturgeon Point, New York (1992–2001) [34], and Great Lakes (1996–2007) [36]

Second, since PCB concentrations are observed to be declining worldwide, we evaluate whether our modeling can reproduce the observed declines in concentrations in a populated region [34–36] (Cell 79 of the BETR-Global grid, Fig. 2.2) and three Arctic regions [37] (Cells 8, 13 and 38 of the BETR-Global grid, Fig. 2.2). The declines in concentrations are approximately characterized by apparent first-order half-lives ($t_{1/2,conc}$), which are defined as the time required for atmospheric concentration to drop by 50%. Table 3.1 shows the comparison between modeled and observed $t_{1/2,conc}$, demonstrating that measured $t_{1/2,conc}$ fall generally within the predicted $t_{1/2,conc}$ ranges. That is, our modeling succeeds in capturing the temporal trends in atmospheric PCB concentrations.

3.4   Overview of the Fate of PCBs in the Anthroposphere

Figure 3.2 displays the CiP-CAFE-modeled fate of Σ_6PCBs in the global anthroposphere from 1930 to 2015. Of the 303 kt of Σ_6PCBs produced worldwide during the historical production period (1930–1993), 286 kt (94.5%) eventually entered the service life of products (LC4) while the remaining 5.5% were emitted (0.03%) or wasted (5.47%) during industrial processes (LC1 and LC2). The calculated fraction not in service (5.5%) is analogous to the assumption in Breivik et al. [33, 43] that 5% of annual consumption was lost during industrial activities. However, whereas Breivik et al. attributed the entire loss to direct atmospheric emissions, CiP-CAFE suggests that the major portion is subject to waste disposal first and hence only partially emitted into the environment. Our calculated fraction of emissions (0.03%) agrees with historical estimates that, for example, 1.18×10^6 lb of PCBs in industrial waste were disposed of while only 3.3×10^3 lb were directly discharged into the environment in 1974 alone [44].

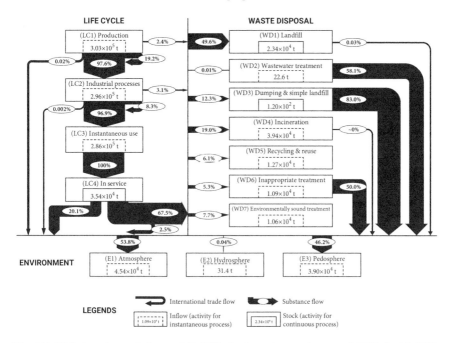

Fig. 3.2 Global total mass balance of Σ_6PCBs in the anthroposphere as of 2015. In this PCB-specific case, inappropriate treatment (WD6) refers to the open burning of PCB-containing waste, and environmentally sound treatment (WD7) refers to complete destruction or irreversible transformation of PCBs (emissions are negligible). All outflows from a process are normalized to 100% and indicated as percentages in ovals. Adapted and reproduced with permission (Li and Wania [32]) Copyright (2016) Elsevier

At the end of 2015, 12.4% (35.4 kt) of the cumulative historical usage remained in in-use stocks (LC4) while 67.5% had entered the waste stream (Fig. 3.2). On a global scale, landfill received the dominant share (WD1: 103 kt; 49.6%) of PCB-containing waste, followed by incineration (WD4: 39.4 kt; 19.0%) and dumping and simple landfill (WD3: 25.5 kt; 12.3%). The latter (WD3) is identified as the most problematic disposal option since it liberated the highest fraction (83.0%) of Σ_6PCBs into the environment; this fraction is even higher than that released from inappropriate treatment (WD6; 50.0%). Despite this slightly lower release fraction, open burning (WD6) is deemed a dangerous waste disposal option because PCBs serve as precursors in the formation of more toxic compounds like polychlorinated dibenzofurans [45, 46]. Compared with dumping and simple landfill (WD3), engineered landfill (WD1) is calculated to release a small fraction (0.03%) of Σ_6PCBs because of the release reduction by landfill gas and leachate collection systems as well as the sequestration of PCB-containing residuals in secluded engineered pits.

Furthermore, Fig. 3.2 shows that atmosphere (E1) and pedosphere (E3) are the two major environmental compartments receiving PCB emissions from the anthroposphere. Figure 3.3 presents the time course of Σ_6PCB emissions worldwide and the contributions from individual anthropospheric sources, i.e., industrial activities (LCs 1 and 2), the use phase (LC 4), and disposal of industrial (waste arising from LCs 1 and 2) and end-of-life waste (waste generated after LC4). Overall, the temporal trend of annual emissions follows an approximately right-skewed bell curve, which peaked in the early 1970s but extends to almost 2080 if no additional regulatory measures are taken in the future.

Of the cumulative emissions of Σ_6PCBs worldwide for the simulation period 1930–2100, passive volatilization from the use phase is estimated to contribute 66.3% (60.6 kt) and the disposal of the end-of-life waste another 31.4% (28.7 kt). Only a tiny fraction (2.3%; 2.1 kt) is generated during industrial activities and treatment of industrial waste (Fig. 3.3). Therefore, our calculation highlights the significance of emissions from the in-use and end-of-life waste stocks. The minor importance of industrial activities explains the previous success [47, 48] in predicting the environmental occurrence of PCBs using the emission estimates by Breivik et al. that neglected industrial emissions and focused on usage and disposal sources alone. Such a modeled source pattern also confirms an earlier conclusion that passive volatilization is the major pathway by which PCBs enter the environment. In particular, it implies that temperature change, e.g., either under a climate change scenario [47] or due to seasonal or diel variation [38, 49], can be an important factor governing PCB emissions. As such, it would be of interest to explore in future studies the importance of this temperature effect for a wider range of CiPs beyond PCBs, which has been argued to "depend on the relative importance of different pathways throughout the chemical life cycle" and thus does "not lend itself well to simple generalizations" [50]. Fortunately, since CiP-CAFE links emissions with their sources and pathways

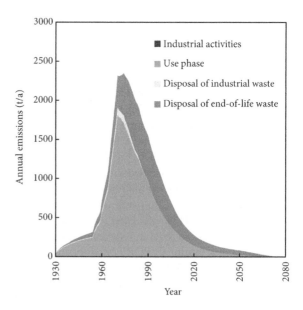

Fig. 3.3 Temporal evolution of annual emissions of Σ_6PCBs and the relative contribution of lifecycle sources. Adapted and reproduced with permission (Li and Wania [32]) Copyright (2016) Elsevier

in an explicit, mechanistic manner, it provides unique opportunities for future studies systematically quantifying this effect of temperature on relevant parameters (e.g., emission factors, the depleting rates of in-use and waste stocks) and hence the long-term emissions.

Since the in-use and waste stocks are identified as the most relevant sources of PCB emissions, one would be curious about the temporal evolution of the two stocks worldwide. Figure 3.4 shows a breakdown of the time-dependent fate of Σ_6PCBs. Continuous use of PCB-containing products led to an accumulation of Σ_6PCBs in in-use stocks, while continuous obsoleting and discarding led to their depletion: these two conflicting driving forces became comparable in 1980 when global in-use Σ_6PCBs stocks peaked at 150 kt (Fig. 3.4). At that time, most in-use PCBs resided in pole-mounted large transformers (AP4, 56.7%), followed by nominally closed systems (AP3, 22.0%) and open usage (AP1, 13.4%) (data not shown in Fig. 3.4). Therefore, eliminating PCBs from large transformers should be given priority in national or regional PCB management strategies. At the end of the product lifecycle, Σ_6PCBs are anticipated to flow from the in-use stocks into waste stocks (Fig. 3.4a). By 2015, on a global scale, 183 kt (60.5% of the cumulative historical production) of Σ_6PCBs had migrated into waste stocks (including both industrial and end-of-life waste), while 35.4 kt (11.7%) were still in use (Fig. 3.4b).

From an industrial ecological point of view, Fig. 3.4b also indicates that both in-use and waste stocks are temporary sinks, in which PCBs are stored for a limited period. By contrast, the environment (receiving 30.2% of total Σ_6PCBs by 2100) and decomposition dissipation (receiving 68.8% of total Σ_6PCBs by 2100) are "final" sinks because PCBs are held for a long time or completely destroyed in these phases and no longer return to the anthroposphere [51]. However, the environment is not a safe final sink [51], as emissions into the environment have the potential to cause

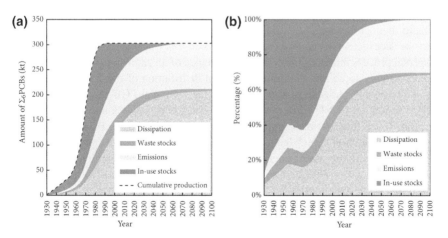

Fig. 3.4 Temporal evolution of global stocks, emissions and decomposition dissipation of Σ_6PCBs in absolute (**a**) and relative (**b**) terms. Reproduced with permission (Li and Wania [32]) Copyright (2016) Elsevier

adverse effects on the ecosystem and humans. Material management strategies should seek to maximize the substance flows to decomposition dissipation and minimize environmental emissions.

3.5 Dispersal of PCBs in the Global Anthroposphere and Environment

The combination of (i) the international trade flows in the anthroposphere and (ii) natural movements in the environment gradually disperse the cumulative Σ_6PCB burdens (mass in a location) from historical manufacturing regions to almost all corners of the world. Figure 3.5 shows the stepwise dispersal of PCBs worldwide. At the end of 2015, 87.6 kt of Σ_6PCBs had left the region of origin and traveled elsewhere via the international trade of substance (after LC1; 55.6 kt, 63.7%), PCB-containing capacitors and transformers (after LC2; 24.4 kt, 27.9%), as well as WEEE (after LC4; 7.3 kt, 8.4%).

Note that these three international trade flows can involve the transport of the same chemical in either direction. For instance, historical records suggest that the United States (Region 6) had purchased "a considerable proportion" of technical PCB substance from Japan (Region 2) [52] but meanwhile exported back to Japan electric equipment [53] which might involve a certain amount of PCBs therein.

The visualization in Fig. 3.5 demonstrates that the geographic distribution of the cumulative Σ_6PCB burden (the mass in a grid cell) becomes increasingly uniform in space as PCBs are spread out. In other words, the global dispersal of PCBs lowers the "non-uniformity" of the cumulative Σ_6PCB burden. To quantify such non-uniformity, we use a coefficient of non-uniformity (NU), which is defined as the ratio of the square root of the variances of Σ_6PCBs burdens in individual grid cells to their global mean (similar to the definition of the "relative standard deviation" or "coefficient of variation" in statistics). In this way, the reduction in NU by a dispersal process indicates the relative importance of an anthropospheric or environmental driving force in smoothing the global distribution of the Σ_6PCBs burdens. As shown in Fig. 3.5, international trade of the technical PCB mixture causes the largest reduction in NU (difference of 503% between panels a and b of Fig. 3.5), i.e. constitutes the dominant anthropogenic force dispersing PCBs throughout the world. The reduction in NU due to international trade of capacitors and transformers (difference of 40% between panels b and c) is almost comparable with natural driving forces (difference of 76% between panels d and e) such as atmospheric advection, which has been described as "the most efficient long-range transport medium" for persistent organic pollutants [54]. This analysis supports the argument that products are "important vehicles of the global transport of chemicals" [55]. While at a global scale the product-related transport comes second to transport by chemical trade, at a regional scale it can be the predominant vehicle. For instance, a previous survey indicated that import of PCB-containing electric equipment is responsible for 33%

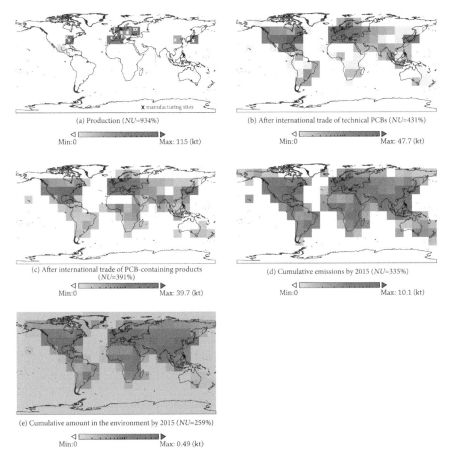

(a) Production (NU=934%)

Min:0   Max: 115 (kt)

(b) After international trade of technical PCBs (NU=431%)

Min:0   Max: 47.7 (kt)

(c) After international trade of PCB-containing products
(NU=391%)

Min:0   Max: 39.7 (kt)

(d) Cumulative emissions by 2015 (NU=335%)

Min:0   Max: 10.1 (kt)

(e) Cumulative amount in the environment by 2015 (NU=259%)

Min:0   Max: 0.49 (kt)

Fig. 3.5 Maps indicating the global spatial dispersal of Σ_6PCBs (on a logarithmic scale) due to different anthropospheric and environmental driving forces. The extent of non-uniformity in each map is expressed with a coefficient of non-uniformity (NU). Adapted and reproduced with permission (Li and Wania [32]) Copyright (2016) Elsevier

(4.5 kt) of total historical PCB usage (13.5 kt) in China [56], while there has never been deliberate import of PCBs as chemicals [57]. CiP-CAFE supports this point by calculating that 52% (1.2 kt) of Σ_6PCB used (2.3 kt) in China (Region 1) entered the country with PCB-containing equipment. The NU difference between panels c and d (56%) is a consequence of different average lifetimes of PCB-containing products, various regional waste disposal patterns and international trade of WEEE; however, it is difficult to discriminate the contribution of each of these driving forces.

As indicated above, the international trade of WEEE is identified as the small-est (8.4%) of the international trade flows of Σ_6PCB; this explains the finding in Breivik et al. [33] that WEEE-related emissions "have only a modest impact" on the PCB contamination worldwide. Despite its minor share at a global scale, the international trade of WEEE is responsible for the high levels of PCB contamination currently occurring in developing regions such as China (Region 1) and South Asian countries (Region 3) [58]. For instance, our calculations suggest that the import of Σ_6PCBs with WEEE (5.3 kt) is 1.8 times higher than the cumulative historical usage of Σ_6PCBs (2.9 kt) in mainland China (Region 1) by 2015. In general, the translo-cation of WEEE from developed to developing regions amplifies emissions, because developing countries (i) often depend more on primitive disposal options such as dumping and simple landfill (WD3) and inappropriate treatment (WD6) [59]; (ii) are mostly distributed in subtropical and tropical regions, in which warm temperatures enhance atmospheric emissions that are mainly governed by passive volatilization [58].

3.6 Summary

The combined anthropospheric and environmental fate modeling in this chapter pro-vide a comprehensive overview of the temporal evolution of stocks and flows of PCBs in the global anthroposphere. We reveal that in-use and waste stocks are tem-porary sinks for PCBs in the anthroposphere, as well as relevant sources of PCBs to the global physical environment. After a sufficiently long period of time, PCBs either permanently disappear from lifecycle and waste disposal phases or enter the physical environment. In the meanwhile, PCBs can undergo global dispersal through both international trades in human society and long-range atmospheric or oceanic movements in the natural environment. The extent of global dispersal caused by humans is larger than that occurring in the natural environment. The role of the international trade of technical substances is more prominent than that of other trade flows. While the international trade of waste has only a modest impact on the global PCB contamination, it does substantially influence the PCB contamination in specific regions.

References

1. Breivik K, Sweetman A, Pacyna JM, Jones KC (2007) Towards a global historical emission inventory for selected PCB congeners—a mass balance approach: 3. an update. Sci Total Environ 377(2):296–307
2. Meijer SN, Ockenden WA, Sweetman A, Breivik K, Grimalt JO, Jones KC (2003) Global distribution and budget of PCBs and HCB in background surface soils: implications for sources and environmental processes. Environ Sci Technol 37(4):667–672

3. Risebrough R, Rieche P (1968) Polychlorinated biphenyls in the global ecosystem. Nature 220:1098–1102
4. Wassermann M, Wassermann D, Cucos S, Miller HJ (1979) World PCBs map: storage and effects in man and his biologic environment in the 1970s. Ann N Y Acad Sci 320(1):69–124
5. Pozo K, Harner T, Wania F, Muir DCG, Jones KC, Barrie LA (2006) Toward a global network for persistent organic pollutants in air: results from the GAPS study. Environ Sci Technol 40(16):4867–4873
6. McCarty JP, Hoffman D, Rattner B, Burton G, Cairns J (2003) The Hudson river: PCB case study. In: Hoffman DJ, Rattner BA, Burton GJA, Cairns JJ (eds) Handbook of ecotoxicology, pp 813–831
7. Hermanson MH, Johnson GW (2007) Polychlorinated biphenyls in tree bark near a former manufacturing plant in Anniston, Alabama. Chemosphere 68(1):191–198
8. Hermanson MH, Scholten CA, Compher K (2003) Variable air temperature response of gas-phase atmospheric polychlorinated biphenyls near a former manufacturing facility. Environ Sci Technol 37(18):4038–4042
9. Diamond ML, Melymuk L, Csiszar SA, Robson M (2010) Estimation of PCB stocks, emissions, and urban fate: will our policies reduce concentrations and exposure? Environ Sci Technol 44(8):2777–2783
10. Shanahan CE, Spak SN, Martinez A, Hornbuckle KC (2015) Inventory of PCBs in Chicago and opportunities for reduction in airborne emissions and human exposure. Environ Sci Technol 49(23):13878–13888
11. Gioia R, Akindele AJ, Adebusoye SA, Asante KA, Tanabe S, Buekens A, Sasco AJ (2014) Polychlorinated biphenyls (PCBs) in Africa: a review of environmental levels. Environ Sci Pollut Res 21(10):6278–6289
12. Wania F, Su Y (2004) Quantifying the global fractionation of polychlorinated biphenyls. Ambio 33(3):161–168
13. Wania F (1998) The significance of long range transport of persistent organic pollutants by migratory animals (WECC Report 3/98). WECC-Report, vol 3. WECC Wania Environmental Chemists Corp.
14. Li N, Wania F, Lei YD, Daly GL (2003) A comprehensive and critical compilation, evaluation, and selection of physical-chemical property data for selected polychlorinated biphenyls. J Phys Chem Ref Data 32(4):1545–1590
15. MacLeod M, Scheringer M, Hungerbühler K (2007) Estimating enthalpy of vaporization from vapor pressure using Trouton's rule. Environ Sci Technol 41(8):2827–2832
16. Sinkkonen S, Paasivirta J (2000) Degradation half-life times of PCDDs, PCDFs and PCBs for environmental fate modeling. Chemosphere 40(9):943–949
17. Cui S, Fu Q, Ma W, Song W, Liu L, Li Y-F (2015) A preliminary compilation and evaluation of a comprehensive emission inventory for polychlorinated biphenyls in China. Sci Total Environ 533:247–255
18. Zhao S, Breivik K, Liu G, Zheng M, Jones KC, Sweetman AJ (2017) Long-term temporal trends of polychlorinated biphenyls and their controlling sources in China. Environ Sci Technol 51(5):2838–2845
19. Song S, Xue J, Lu Y, Zhang H, Wang C, Cao X, Li Q (2018) Are unintentionally produced polychlorinated biphenyls the main source of polychlorinated biphenyl occurrence in soils? Environ Pollut 243:492–500
20. Breivik K, Sweetman A, Pacyna JM, Jones KC (2002) Towards a global historical emission inventory for selected PCB congeners—a mass balance approach: 2. emissions. Sci Total Environ 290(1):199–224
21. Allende D, Ruggeri MF, Lana B, Garro K, Altamirano J, Puliafito E (2016) Inventory of primary emissions of selected persistent organic pollutants to the atmosphere in the area of Great Mendoza. Emerg Contam 2(1):14–25
22. Zhang G, Li X, Mai B, Peng P, Ran Y, Wang X, Zeng EY, Sheng G, Fu J (2007) Sources and occurrence of persistent organic pollutants in the pearl river delta, south China. In: Li A, Tanabe S, Jiang G, Giesy JP, Lam PKS (eds) Developments in environmental science, vol 7. Elsevier, Amsterdam and Oxford, pp 289–311

23. Robson M, Melymuk L, Csiszar SA, Giang A, Diamond ML, Helm PA (2010) Continuing sources of PCBs: the significance of building sealants. Environ Int 36(6):506–513
24. Diefenbacher PS, Gerecke AC, Bogdal C, Hungerbühler K (2016) Spatial distribution of atmospheric PCBs in Zurich, Switzerland: do joint sealants still matter? Environ Sci Technol 50(1):232–239
25. Kohler M, Tremp J, Zennegg M, Seiler C, Minder-Kohler S, Beck M, Lienemann P, Wegmann L, Schmid P (2005) Joint sealants: an overlooked diffuse source of polychlorinated biphenyls in buildings. Environ Sci Technol 39(7):1967–1973
26. Breivik K, Sweetman A, Pacyna JM, Jones KC (2002) Towards a global historical emission inventory for selected PCB congeners—a mass balance approach: 1. global production and consumption. Sci Total Environ 290(1):181–198
27. United Nations Commodity Trade Statistics Database (Comtrade) (2017)
28. Statistics of US Business (SUSB) (2008–2012 Annual Datasets) (2017)
29. United Nations Industrial Commodity Statistics Database (2017) United Nations Statistics Division
30. China National Chemical Information Center (1997–2016) China Chemical Industry Yearbook 1997–2016. China Petroleum & Chemical Industry Federation, Beijing
31. Breivik K, Armitage JM, Wania F, Jones KC (2014) Tracking the global generation and exports of e-waste. Do existing estimates add up? Environ Sci Technol 48(15):8735–8743
32. Li L, Wania F (2016) Tracking chemicals in products around the world: introduction of a dynamic substance flow analysis model and application to PCBs. Environ Int 94:674–686
33. Breivik K, Armitage JM, Wania F, Sweetman AJ, Jones KC (2016) Tracking the global distribution of persistent organic pollutants accounting for e-waste exports to developing regions. Environ Sci Technol 50(2):798–805
34. Buehler SS, Basu I, Hites RA (2004) Causes of variability in pesticide and PCB concentrations in air near the Great Lakes. Environ Sci Technol 38(2):414–422
35. Sun P, Basu I, Hites RA (2006) Temporal trends of polychlorinated biphenyls in precipitation and air at Chicago. Environ Sci Technol 40(4):1178–1183
36. Venier M, Hites RA (2010) Time trend analysis of atmospheric POPs concentrations in the Great Lakes Region since 1990. Environ Sci Technol 44(21):8050–8055
37. Hung H, Katsoyiannis AA, Brorström-Lundén E, Olafsdottir K, Aas W, Breivik K, Bohlin-Nizzetto P, Sigurdsson A, Hakola H, Bossi R, Skov H, Sverko E, Barresi E, Fellin P, Wilson S (2016) Temporal trends of persistent organic pollutants (POPs) in Arctic air: 20 years of monitoring under the Arctic monitoring and assessment programme (AMAP). Environ Pollut 217:52–61
38. Bogdal C, Müller CE, Buser AM, Wang Z, Scheringer M, Gerecke AC, Schmid P, Zennegg M, MacLeod M, Hungerbühler K (2014) Emissions of polychlorinated biphenyls, polychlorinated dibenzo-p-dioxins, and polychlorinated dibenzofurans during 2010 and 2011 in Zurich, Switzerland. Environ Sci Technol 48(1):482–490
39. Halse AK, Schlabach M, Eckhardt S, Sweetman A, Jones KC, Breivik K (2011) Spatial variability of POPs in European background air. Atmos Chem Phys 11(4):1549–1564
40. Jaward FM, Farrar NJ, Harner T, Sweetman AJ, Jones KC (2004) Passive air sampling of PCBs, PBDEs, and organochlorine pesticides across Europe. Environ Sci Technol 38(1):34–41
41. Shahpoury P, Lammel G, Holubová Šmejkalová A, Klánová J, Přibylová P, Váňa M (2015) Polycyclic aromatic hydrocarbons, polychlorinated biphenyls, and chlorinated pesticides in background air in central Europe—investigating parameters affecting wet scavenging of polycyclic aromatic hydrocarbons. Atmos Chem Phys 15(4):1795–1805
42. Teil M-J, Blanchard M, Chevreuil M (2004) Atmospheric deposition of organochlorines (PCBs and pesticides) in northern France. Chemosphere 55(4):501–514
43. Breivik K, Czub G, McLachlan MS, Wania F (2010) Towards an understanding of the link between environmental emissions and human body burdens of PCBs using CoZMoMAN. Environ Int 36(1):85–91
44. USEPA (1976) PCBs in the United States: industrial use and environmental distribution (task I). U.S. Environmental Protection Agency, Washington, DC

45. Buser HR, Bosshardt H-P, Rappe C (1978) Formation of polychlorinated dibenzofurans (PCDFs) from the pyrolysis of PCBs. Chemosphere 7(1):109–119

46. Buser HR (1979) Formation of polychlorinated dibenzofurans (PCDFs) from the pyrolysis of individual PCB isomers. Chemosphere 8(3):157–174

47. Lamon L, von Waldow H, MacLeod M, Scheringer M, Marcomini A, Hungerbühler K (2009) Modeling the global levels and distribution of polychlorinated biphenyls in air under a climate change scenario. Environ Sci Technol 43(15):5818–5824

48. MacLeod M, Riley WJ, Mckone TE (2005) Assessing the influence of climate variability on atmospheric concentrations of polychlorinated biphenyls using a global-scale mass balance model (BETR-Global). Environ Sci Technol 39(17):6749–6756

49. Gasic B, Moeckel C, MacLeod M, Brunner J, Scheringer M, Jones KC, Hungerbühler K (2009) Measuring and modeling short-term variability of PCBs in air and characterization of urban source strength in Zurich, Switzerland. Environ Sci Technol 43(3):769–776

50. Gouin T, Armitage JM, Cousins IT, Muir DC, Ng CA, Reid L, Tao S (2013) Influence of global climate change on chemical fate and bioaccumulation: the role of multimedia models. Environ Toxicol Chem 32(1):20–31

51. Kral U, Kellner K, Brunner PH (2013) Sustainable resource use requires "clean cycles" and safe "final sinks". Sci Total Environ 461:819–822

52. de Voogt P, Brinkman UAT (1989) Production, properties and usage of polychlorinated biphenyls. In: KR D, Jensen AA (eds) Halogenated biphenyls, terphenyls, naphthalenes, dibenzodioxins and related products. Elsevier Science Publishers B.V., Amsterdam

53. Yanari T (2004) History of power transformers in Japan and description of historical materials. Survey reports on the systemization of technologies no. 4. Center of the History of Japanese Industrial Technology, National Museum of Nature and Science, Tokyo

54. Lohmann R, Breivik K, Dachs J, Muir D (2007) Global fate of POPs: current and future research directions. Environ Pollut 150(1):150–165

55. UNEP (2019) Global chemicals outlook II. From legacies to innovative solutions: implementing the 2030 agenda for sustainable development. United Nations Environment Programme, Nairobi

56. World Bank (2005) Project document on a proposed grant from the global environment facility trust fund in the amount of USD 18.34 Million to the People's Republic of China for a PCB management and disposal demonstration project. Environment Sector Unit (EASEN) East Asia and Pacific Region, Bangkok

57. People's Republic of China (2007) The People's Republic of China national implementation plan for the stockholm convention on persistent organic pollutants. The Government of People's Republic of China, Beijing

58. Breivik K, Gioia R, Chakraborty P, Zhang G, Jones KC (2011) Are reductions in industrial organic contaminants emissions in rich countries achieved partly by export of toxic wastes? Environ Sci Technol 45(21):9154–9160

59. Weber R, Watson A, Forter M, Oliaei F (2011) Review article: persistent organic pollutants and landfills-a review of past experiences and future challenges. Waste Manag Res 29(1):107–121

Chapter 4
The Degradation of Fluorotelomer-Based Polymers Contributes to the Global Occurrence of Fluorotelomer Alcohols and Perfluoroalkyl Carboxylates

Abstract In this chapter, we investigated whether the degradation of side-chain fluorotelomer-based polymers (FTPs), mostly in waste stocks (i.e., landfills and dumps), serves as a long-term source of fluorotelomer alcohols (FTOHs) and perfluoroalkyl carboxylates (PFCAs) to the global environment. Our modeling results indicate that the estimates of in-use and waste stocks of FTPs are more sensitive to the selected lifespan of finished products, while those of the emissions of FTOHs and PFCAs are more sensitive to the selected half-life for degradation of FTPs in waste stocks. FTP degradation in waste stocks is making an increasing contribution to FTOH generation worldwide, the bulk of which readily migrates from waste stocks and is degraded into PFCAs in the environment. The remaining part of the generated FTOHs is degraded in waste stocks, which makes those stocks reservoirs that slowly release PFCAs into the environment over the long run because of the low leaching rate and extreme persistence of PFCAs.

4.1 Introduction

Constituting almost 80% of the market of fluorotelomer-based substances worldwide [1], side-chain fluorotelomer-based polymers (FTPs) have been applied (i) as durable water repellents (DWRs) on a wide range of finished textiles, fabrics, carpets and garments, (ii) as oil and grease repellents in paper and packaging industries, and (iii) to other miscellaneous applications [2]. The FTPs provide continuous water, oil and stain resistance for commercial finished products throughout the product lifespans. At the end of product lifespans, FTPs enter the waste stream and accumulate in waste stocks such as landfills and dumps, where aged FTPs undergo degradation on the time scale of decades or longer to generate various per- and polyfluoroalkyl substances (PFASs) [3]. This process comprises a series of sequential stepwise transformations, which first form non-polymeric fluorotelomer-based substances like fluorotelomer alcohols (FTOHs), followed by a variety of immediate degradation products such as saturated and unsaturated fluorotelomer carboxylates, and arrive at perfluoroalkyl carboxylates (PFCAs) as ultimate degradation products [4–6].

© Springer Nature Singapore Pte Ltd. 2020
L. Li, *Modeling the Fate of Chemicals in Products*, Springer Theses,
https://doi.org/10.1007/978-981-15-0579-9_4

Despite the consensus that FTP degradation contributes to the occurrence of FTOHs and PFCAs worldwide, the magnitude and temporal evolution of the problem have not yet been well elucidated and are thus still being contested [7]. Specifically, the academic attempt to address the following questions:

(i)   How much FTOHs and PFCAs are released into the environment worldwide due to the stepwise degradation of FTPs? How do the annual releases evolve with time, increasing or decreasing?

(ii) How do the lifespan of finished products [hereafter "product lifespan (LS)"], and the degradation half-life of FTPs in the environment or waste stocks [hereafter "degradation half-life (HL)"], influence the degradation of FTPs and annual releases of FTOHs and PFCAs?

(iii) Where does the degradation mainly occur, in waste stock of the anthroposphere or in the environment?

To answer these questions, we combine anthropospheric (CiP-CAFE) and environmental (BETR-Global) fate models to mechanistically simulate the temporal evolution of in-use and waste stocks of FTPs, and the environmental releases of FTOHs and PFCAs from the degradation of FTPs. The influence of variations in LSs and HLs on model predictions is explored with four scenarios. We seek to preliminarily characterize the current and future contributions of FTP degradation, in particular during the waste disposal phase, to the releases of FTOHs and PFCAs worldwide.

4.2   Method and Data

4.2.1   Applications, Chemicals and Terminology

Since terminology varies substantially between earlier researches and publications, we first harmonize the terminology used throughout this chapter.

Overall, PFASs can be found in products as (i) *ingredients*, i.e., intentional components; (ii) *impurities*, i.e., undesired byproducts; and (iii) *residuals*, i.e., unreacted raw materials. In addition, a PFAS can be a *degradation product* of other PFASs.

In terms of fluorotelomer-based substances, main applications (APs) include [8, 9]: (i) the use of *polymeric* fluorotelomer-based substances as DWRs on finished textiles, fabrics, carpets and garments (AP1), and (ii) the use of *non-polymeric* fluorotelomer-based derivatives as surfactants for treating consumer products (representing all uses with continuous releases throughout lifespan) (AP2) and additives in aqueous film forming foams (AFFFs, representing all uses with both accidental releases at accidents and intensive discharge at the end of shelf life) (AP3).

Since technical PFASs are usually present and consumed in the above applications as mixtures of homologues with different chain-lengths or derivatives with various functional groups (e.g., sulfonamide, betaine) or cations, we defined "equivalents" to collectively describe a series of similar PFASs with the same featured moiety but without considering their differences in molecular weights, physicochemical properties and degradation kinetics. The following three categories of PFAS-equivalents will be considered in this chapter (Fig. 4.1):

(i) side-chain FTP-equivalents (respectively, 4:2, 6:2, 8:2, 10:2 and 12:2 homologues), which refer to a collection of side-chain fluorotelomer-based acrylate, methacrylate, urethane and other polymers. FTP-equivalents are used as *ingredients* in DWRs in AP1 (denoted as FTPs) and have the potential to be degraded into FTOHs and further PFCAs;

(ii) FTOH-equivalents (respectively, 4:2, 6:2, 8:2, 10:2 and 12:2 homologues), which refer to a collection of (i) FTOHs, and (ii) fluorotelomer-derived non-polymers (e.g., fluorotelomer acrylates), the degradation of which generates FTOHs as intermediates [10]. FTOH-equivalents can be released into the environment as *residuals* in consumer products in the three APs (denoted as res-FTOHs, Fig. 4.1), and as *degradation products* from the degradation of FTPs (denoted as degFTOHs, Fig. 4.1) in both waste stocks and the environment;

(iii) PFCA-equivalents (respectively, C4–C12 homologues), which refer to both perfluoroalkyl carboxylic acids and corresponding carboxylates. In this study,

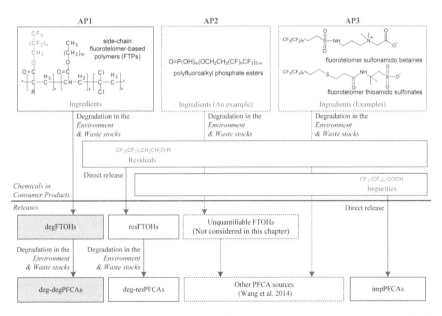

Fig. 4.1 Transformation of FTPs, FTOHs and PFCAs in the environment and waste stocks [R'=H or methyl group; R=H or (meth)acrylic group]. Arrows in solid lines denote the substance flows quantified in this work: estimating stocks of FTPs, and annual releases of degFTOHs and deg-degPFCAs, are the main objectives (denoted as blue and red shadings); annual releases of resFTOHs, deg-resPFCA and impPFCAs are also calculated for comparison. Arrows in dashed lines represent substance flows not quantified. The box "other PFCA sources" includes all PFCAs sources in Wang et al. (2014) [9] other than the three considered here. Reproduced with permission (Li et al. [24]) Copyright (2017) American Chemical Society

we consider releases of PFCA-equivalents into the environment as *impurities* in consumer products in the three APs (denoted as impPFCAs, Fig. 4.1), and as *degradation products* from both resFTOHs (denoted as deg-resPFCAs, Fig. 4.1) and degFTOHs (denoted as deg-degPFCAs, Fig. 4.1) in waste stocks and the environment.

For each FTOH and neutral PFCA homologue, the partitioning coefficients (K_{AW}, K_{OW} and K_{OA}) at 25 °C are taken from Kim et al. [11], which have been adjusted for thermodynamic consistency. Since the air-water partitioning of PFCAs depends on the speciation between their acid and conjugate base forms, we adopted the "D_{AW}" which related the K_{AW} values for neutral molecules and the distribution ratio between anion and acid. The derivation of effective D_{AW} follows the method by Armitage et al. [12]; the acid dissociation constants (pK_a) of individual PFCA homologues range from 0.7 to 0.8 [13]. Internal energies of phase transfer (ΔU_W, ΔU_A and ΔU_O), which are employed to adjust the partition coefficients for temperature dependence, are calculated following MacLeod et al. [14] based on Trouton's Rule and defaults. We collect measured or assumed degradation half-lives of FTOHs in air [15], water (and wastewater) [16, 17], soil, and sediment (and solid waste in landfill, and dumping and simple landfill) [18–20], and degradation half-lives of PFCAs in air [21], water (and wastewater) [22], soil, and sediment (and solid waste in landfill, and dumping and simple landfill) [23]. Activation energies for adjusting the degradation half-lives for temperature dependence are defaults in the CiP-CAFE and BETR-Global models.

4.2.2   Modeling Strategy

4.2.2.1   Step 1: Modeling the Time-Dependent Fate of FTPs

The first step of our modeling is to perform CiP-CAFE calculations to obtain the time-variant estimates of the global in-use and waste (landfill + dumping and simple landfill) stocks of FTPs and their release for the period 1960–2040.

We first derive the annual production of FTPs on a homologue-basis in individual CiP-CAFE regions. The global annual production of total technical fluorotelomer-based substances [8, 9, 25] is split into FTPs (for use in AP1) and non-polymeric substances (for use in APs2 and 3) based on a reported ratio of 80%:20% [1]. Furthermore, because technical fluorotelomer-based substances are mixtures of different homologues of ingredients, residuals and impurities, we divide these mixtures according to literature-reported homologue composition (for ingredients) and contents (for residuals and impurities) in long-chain (C8-based) and short-chain (C6-based) products (for details, see Li et al. [24]). To reflect a worldwide transition from long-chain (C8-based) products to their short-chain (C6-based) alternatives since 2006, we assume that the fraction of long-chain products in total fluorotelomer-based products decreased linearly from 100% before 2005 to 0% after 2016. This simplified assumption could underestimate the amounts of long-chain products in developing countries (e.g., in China [25]) as domestic production and use of long-chain products are still ongoing in these countries. Our method for estimating the global annual production is believed to be reliable because, for example, the estimate for C8 FTPs for the period 1970–2007 (41.4–49.7 kilotonnes, kt) compares favorably with a previous report of 34.5 kt (32 kt of acrylate [4] plus 2.5 kt of urethane polymers [26]).

Next, the global annual production of FTPs on a homologue-basis is attributed into the regions defined in CiP-CAFE: ~5% in Western Europe (Region 5) [27], 50% in North America (Region 6) [28], ~15% in mainland China (Region 1, only after 2009) [25], and the remainder in Japan/South Korea (Region 2) (for details, see Li et al. [24]).

As indicated above, FTPs are used as DWRs on finished textiles, fabrics, carpets and garments (AP1). Recalling that our goal is to investigate the influence of different product lifespans (LSs) in AP1 and degradation half-lives (HLs) of FTPs in waste stocks on our model results, we run CiP-CAFE with four scenarios: (I) LS = 10 years and HL = 75 years, (II) LS = 10 years and HL = 1500 years, (III) LS = 50 years and HL = 75 years, and (IV) LS = 50 years and HL = 1500 years.

CiP-CAFE sketches interregional mass exchange of FTPs through trade flows of (i) concentrated DWRs (after LC1) and (ii) finished products such as textiles, fabrics, carpets and garments (after LC2). We use concentrated DWRs as a surrogate for the FTPs subject to textile/fabric finishing after the "production (LC1)" stage, and clothing as a surrogate for the FTPs in finished textile/fabric products reaching consumers after the "industrial processes (LC2)" stage. International trade information for the concentrated DWRs and clothing is compiled from the literature (for details, see Li et al. [24]). Due to a lack of temporally resolved data, import and export fractions of the concentrated DWRs and clothing are assumed to be constant throughout the modeled period.

Emission and waste factors for lifecycle stages are given in Li et al. [24]. Here, we assumed that FTPs are not degraded (decomposed) during the product lifespan, as the literature shows a substantial difference in the observed reduction in oil and water repellency or FTP weight between diverse (co)polymer compositions, finishing treatments, and textile surface properties in laundering and/or weathering tests [29–31]. Emissions of FTPs from waste stocks are assumed to be negligible because FTPs are neither volatile nor soluble, whereas degradation of FTPs in waste stocks is calculated using the four assumed HLs listed above.

At the end of their lifespan, obsolete FTP-containing finished products (in solid state) in AP1 enter the waste stream. We assume the absence of inappropriate treatment (WD6) and environmentally sound treatment (WD7) of FTP-containing waste. That is, all FTP-containing waste is distributed among the disposal approaches WDs 1, 3, 4 and 5 following the time-variant regional waste disposal ratios supplied within CiP-CAFE.

4.2.2.2   Step 2: Modeling the Time-Dependent Fate of FTOHs

The second step is composed of another series of CiP-CAFE runs to estimate the
global annual releases of FTOHs for the period 1960–2040. Based on results in
Sect. 4.2.2.1 of (i) FTPs in waste stocks and (ii) FTPs released into the environment,
we calculate the formation of degFTOHs, by conservatively assuming that FTOHs
are formed from FTPs with the same perfluorinated chain length with a yield of
100% [32]. On the other hand, we calculate the substance flow of FTOHs present
as residuals (i.e., resFTOHs) in products in all three APs. The formed degFTOHs
in waste stocks and the annual production of resFTOHs, collectively referred to as
total FTOHs, serve as inputs for the CiP-CAFE calculation of emissions and stocks
of FTOHs. We recognize that a small amount of FTOHs can also be generated from
degradation of nonpolymeric FT-based substances, e.g., polyfluoroalkyl phosphate
esters in AP2 (collectively referred to as "unquantifiable FTOHs" in Fig. 4.1); we
do not consider their contribution to the total FTOH release because (i) inadequate
available market information renders estimating their contribution highly uncertain
and (ii) the release is much lower than degFTOHs and resFTOHs, according to our
preliminary calculation.

The modeling of FTOH fate is also performed on a scenario basis. For product
lifespan in AP1, we use the assumed LSs in the four scenarios in Step 1. Product
lifespan in AP2 is assumed to follow the Weibull distribution with a scale parameter
of 10 years and a shape parameter of 2.4 years, and that in AP3 is associated with
a constant annual loss rate of 4% ± 2% over a shelf life of 15 years. Emission
and waste factors used in the CiP-CAFE calculation are given in Li et al. [24].
Like the calculation for FTPs, we assume that the decomposition of FTOHs during
lifecycle stages is negligible. The waste disposal ratios for waste containing FTOHs
are assumed to be identical to those of waste containing FTPs.

In order to evaluate the CiP-CAFE results for the different scenarios and to calculate FTOH transformation in the environment, we perform global environmental fate
modeling for the total FTOHs (resFTOHs + degFTOHs) using the BETR-Global
model. Time-variant atmospheric concentrations of 8:2 and 6:2 FTOH calculated
by BETR-Global for different scenarios were then compared with monitoring data
reported in the literature.

4.2.2.3   Step 3: Modeling the Time-Dependent Fate of PFCAs

The annually degraded amounts of the total FTOHs (resFTOHs + degFTOHs) in
both waste stocks (calculated by CiP-CAFE in Step 2) and the environment (calculated by BETR-Global in Step 2), are fractionally converted into deg-resPFCAs
and deg-degPFCAs, respectively. The conversion from FTOHs to PFCAs is based
on the median values of the estimated molar yields (mol%) for the transformation
of n:2 FTOH to different PFCA homologues: the degradation of n:2FTOH yields
1 mol% for each PFCA homologue with $(n - 1)$, $(n - 2)$, $(n - 3)$ carbons and

5 mol% for each PFCA homologue with n and $(n + 1)$ carbons [9]. The yields correspond to homologue-specific mass yields of 4.1–6.0% on a FTOH basis, which agree well with those used in previous modeling studies [33–35]. Admittedly, the PFCA yields could be somewhat conservative, because we considered neither degradation intermediates of FTOHs (e.g., fluorotelomer aldehydes [34]), nor the PFCA formation from FTP degradation intermediates other than FTOHs (e.g., newly identified 7:2 sFTOH [F(CF$_2$)$_7$CH(OH)CH$_3$] in bio-degradation of 8:2 fluorotelomer acrylate [10]), due to insufficient information on the multimedia partitioning behavior and degradation kinetics of these intermediates. Meanwhile, we did not consider the PFCA formation in the environment resulting from further transformation of the degradation intermediates released from waste stocks.

The converted deg-resPFCAs and deg-degPFCAs in waste stocks, along with the annual production of impPFCAs in all three APs, served as inputs for a third set of CiP-CAFE calculations to calculate time-variant emissions of total fluoroteolmer-related PFCAs for the period 1960–2040. Again, emission and waste factors used are given in Li et al. [24]. We also assume that the decomposition of PFCAs during lifecycle stages is negligible. The waste disposal ratios for PFCAs are assumed to be identical to those of FTPs.

4.3 Temporal Evolution of Stocks and Releases

Figure 4.2 presents the calculated time-variant global in-use and waste stocks of FTPs, as well as annual releases of degFTOHs and deg-degPFCAs into the environment, for long-chain C8-compounds (panels a to d) and short-chain C6-compounds (panels e to h) under the four scenarios combining LSs and HLs.

We first interpret the in-use and waste stocks of FTPs. Within the given time frame from 1960 to 2040, for both FTPs, scenarios based on a long LS (III and IV) yield larger in-use stocks (Fig. 4.2a and e) but smaller waste stocks (Fig. 4.2b and f), because longer use implies slower transfer to waste. In scenarios with a long HL (II and IV), less FTPs are degraded, resulting in larger waste stocks (Fig. 4.2b and f). For both the in-use and waste stocks, the difference in stock size due to a change in LS (i.e., Scenarios I and II vs. Scenarios III and IV) is more notable than that due to a change in HL (i.e., Scenarios I and III vs. Scenarios II and IV). This indicates that, when both are at possibly realistic levels, product lifespan is more crucial than degradation half-life to an accurate description of FTPs stocks.

In general, both the in-use and waste stocks peak or plateau when inflows and outflows are close to each other in magnitude. For example, in-use stocks of C8 FTPs (Fig. 4.2a) peaked at 20–25 kt in 2010 (Scenarios I and II) or 50–60 kt in 2014 (Scenarios III and IV) when increasing rates of discarding matched decreasing rates of new use (the latter a result of the phase-out of C8 FTPs). Likewise, in-use stocks of C6 FTPs (Fig. 4.2d) are expected to level off at 59–70 kt after 2027 (Scenario I and II) when the increasing discard rates catch up with the annual rates of new

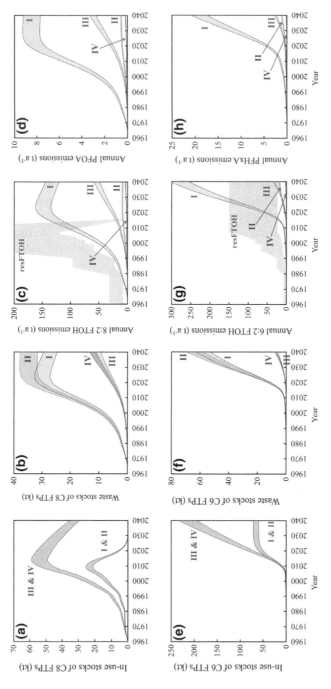

Fig. 4.2 Estimated ranges of global in-use stocks (panels **a** and **e**) and waste stocks (panels **b** and **f**) of FTPs, annual releases of degFTOHs (panels **c** and **g**) and deg-degPFCAs (panels **d** and **h**) for C8- (from **a** to **d**) and C6-compounds (from **e** to **h**) from 1960–2040 under the four simulation scenarios: (I) LS = 10 yrs and HL = 75 yrs, (II) LS = 10 yrs and HL = 1500 yrs, (III) LS = 50 yrs and HL = 75 yrs, and (IV) LS = 50 yrs and HL = 1500 yrs. The global annual releases of resFTOHs are presented in gray shading in panels **c** and **g** for comparison. Reproduced with permission (Li et al. [24]) Copyright (2017) American Chemical Society

use of C6-based finished products (which is assumed constant according to annual production data in Wang et al. [9]).

Next, we discuss the annual releases of degFTOHs (Fig. 4.2c and g). Annual releases of 8:2 degFTOH are anticipated to peak around 2022 in Scenario I and decline afterward, due to the rapid depletion of C8 FTPs waste stocks in this scenario (Fig. 4.2b); while in other scenarios releases of both 8:2 and 6:2 degFTOHs are projected to keep increasing throughout the simulation period. The highest annual releases of 8:2 degFTOH are estimated to range from 2–3 (Scenario IV) to 130–160 t a^{-1} (Scenario I) during the simulation period. Furthermore, for both 6:2 and 8:2 degFTOHs, the annual releases in scenarios with a short HL (I and III) are almost 1 order of magnitude higher than in those using a long HL (II and IV). That is, the estimated annual releases of degFTOHs are affected to a larger extent by a change in HL—which is in contrast to the sensitivity of stocks to LS—when both the HL and LS vary within a realistic range.

Note that degFTOHs here include (i) release of the generated degFTOHs in waste stocks (route 1: released from the anthroposphere), and (ii) transformation from FTPs released into the environment (route 2: formation in the environment). The releases of degFTOHs via the former route vary over three orders of magnitude among the scenarios, with the calculated cumulative releases of 8:2 degFTOHs ranging from 11.5–13.8 t under Scenario IV to 1700–2038 t under Scenario I and those of 6:2 degFTOHs ranging from 0.7–0.9 t under Scenario IV to 126–152 t under Scenario I by 2015. By contrast, the releases of degFTOHs via the latter route were quite similar for the different scenarios, namely 22–37 t for 8:2 degFTOHs and 3–6 t for 6:2 degFTOHs by 2015. Consequently, the relative importance of the two routes varies considerably between the four scenarios.

In addition, we look at the annual releases of deg-degPFCAs (Fig. 4.2d and h). Analogous to the sensitivity of degFTOHs above, HL was identified as the factor with the most influence on the annual releases of deg-degPFCAs, as those releases are considerably higher in Scenarios I and III (Fig. 4.2d and h). Scenario I leads to the highest release of deg-degPFCAs, with a peak in deg-degPFOA of 7–9 t a^{-1} after 2020 (Fig. 4.2d). The annual releases of deg-degPFOA and deg-degPFHxA are predicted to increase throughout the simulation period, except for Scenario I in which releases of deg-degPFOA reach a plateau (Fig. 4.2d).

Again, we should emphasize that deg-degPFCAs combine those (i) liberated from the generated deg-degPFCAs in waste stocks (route 1: released from the anthroposphere) and (ii) transformed from the degFTOHs released into the environment (route 2: formation in the environment). The latter route is currently dominant, with contributions to cumulative releases of 89–95% for deg-degPFOA and 68–97% for deg-degPFHxA by 2015 under different scenarios. Such dominance of the latter route is understandable as our calculations find that (i) 97–99% of degFTOHs volatilize with landfill gas into the atmosphere, where 94–97% of them then degrade into deg-degPFCAs within a year, (ii) by contrast, less than 3% of the deg-degPFCAs generated from the degFTOHs in waste stocks are annually released via leachate. Notably, that leaching is not an efficient route for delivering a significant amount of deg-degPFCAs into the environment in the short term, occurs mostly because the area-specific leachate flow, which equates to the average annual precipitation of ~1 m year^{-1} (the default in CiP-CAFE), is not large. Likewise, Yan et al. [36] estimated that annually 1.8 ± 3 t of PFOA were leached from landfills across China based on samples collected in 2013, which accounts for a mere ~3% of the annual national emissions of PFOA (<60 t in 2012 [25]) in China.

The above result implies that, due to such a "trickle", the residence time of deg-degPFCAs generated in waste stocks can reach multiple decades if not centuries, i.e., waste stocks are a reservoir slowly releasing deg-degPFCAs. In fact, if we extend the calculation to the year 2100, we can find that the releases of deg-degPFCAs from waste stocks will have increased throughout the simulation period (data not shown here), although the formation of deg-degPFCAs in the environment will moderately decline after 2020 due to the depletion of C8 FTP waste stocks (Fig. 4.2b). After 2090, the annual releases from the two routes will be comparable.

4.4 Rising Contributions of FTP Degradation to Global FTOH Releases

In addition to the amount of FTOHs released as a result of FTP degradation (degFTOHs), the CiP-CAFE model also estimates the amount of FTOHs released as residuals (resFTOHs, gray shading in Fig. 4.2c and g). The annual releases of 8:2 resFTOH are estimated to have substantially increased since the 1990s with a peaked around 2007 (the year after the worldwide transition from long-chain-based commercial products to their short-chain alternatives began); those of 6:2 resFTOH have increased steadily until reaching a plateau in 2007. Our estimates agree with earlier studies. For example, our annual 8:2 resFTOH emissions (80–320 t year^{-1} for the period 2000–2004 on average) are generally similar to those reported by Prevedouros et al. [8] (~100 t year^{-1} for 2002), Wania [33] (100–200 t year^{-1} for 2000–2005) and Schenker et al. [35] (60–155 t year^{-1} for 2000–2005); they are also within the range of ~0.5 to ~300 t yr^{-1} which can be calculated from the data presented in Wang et al. [9].

The relative contributions of degFTOHs and resFTOHs to the total releases change over time. 8:2 resFTOH dominated the annual total emissions of 8:2 FTOH prior to approximately 2010, with its contribution exceeding that of 8:2 degFTOH by a factor of 2 (Scenario I) to 2 orders of magnitude (Scenario IV). Since 2010, the relative importance of 8:2 degFTOH has been increasing in the wake of the phase-out of long-chain products in most world regions. Such a transition from resFTOH-dominant to degFTOH-dominant releases is also obvious for Scenario I of 6:2 FTOH (Fig. 4.2g), in which the annual releases of 6:2 degFTOH would exceed that of 6:2 resFTOH by approximately 2025. The dominance of resFTOHs before around 2010 also provides justification for the good agreement between measured and modeled FTOH concentrations in a series of earlier modeling studies [33, 35, 37], which were based on emission estimates of resFTOHs alone.

Furthermore, the combination of the CiP-CAFE and BETR-Global models predicts the atmospheric concentrations of FTOHs based on the emission estimates of resFTOH and degFTOH under Scenario I. Then, the modeled concentrations of 8:2 and 6:2 FTOHs are plotted against measured concentrations (Fig. 4.3), which had been reported from large-scale sampling cruises or on-land campaigns so as to encompass a wide span of representative BETR-Global cells (for an overview of detailed information on sampling sites, see Li et al. [24]). Cells covering the territory of mainland China were excluded from the comparison because the ongoing

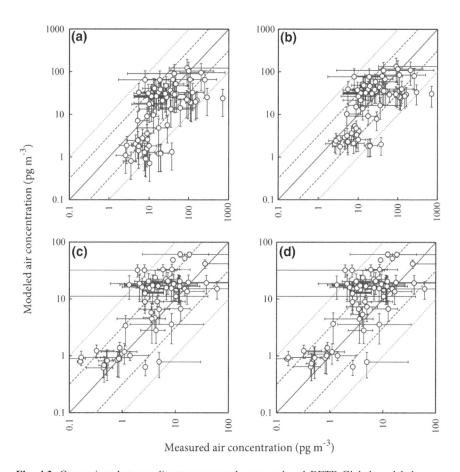

Fig. 4.3 Comparison between literature-reported measured and BETR-Global modeled atmospheric concentrations of 8:2 FTOH (**a** and **b**) and 6:2 FTOH (**c** and **d**) at different sites. The BETR-Global simulations were based on the annual releases of resFTOHs alone (**a** and **c**), and combined resFTOHs and degFTOHs under the Scenario I (**b** and **d**). Diagonal lines represent perfect agreement (solid lines), and agreement within a factor of $10^{0.5}$ (dashed lines) and 10 (dotted lines). Error bars indicate the range of the concentrations (i.e., the difference between maximum and minimum). Reproduced with permission (Li et al. [24]) Copyright (2017) American Chemical Society

production and new uses of long-chain products in China could still be a cause for resFTOH releases thus resulting in an underestimation in 8:2FTOH concentrations (see Sect. 4.2.2.1). Figure 4.3a and b demonstrate good agreement between measured and modeled 8:2 FTOH concentrations, irrespective of whether the releases of 8:2 degFTOH from Scenario I are considered (Fig. 4.2c). However, including releases of 8:2 degFTOH (Fig. 4.3b) does improve model agreement with the more recent measurements, most notably the latest available measurements (at the bottom left) which were sampled between October 2010 and January 2011 [38]. Such an improvement lends supports to our hypothesis of an increasing contribution of degFTOHs. Compared with Scenario I, the improvements in the other scenarios were less remarkable (data not shown here) because the estimated releases of 8:2 degFTOH are much smaller than those of 8:2 resFTOH (Fig. 4.2c). For 6:2 FTOH (Fig. 4.3c and d), including 6:2 degFTOH releases improves model agreement with the more recent measurements as well. However, the improvement is too minor to be readily perceived from the figure, because the estimated releases of 6:2 degFTOH are nearly 2 orders of magnitude lower than those of 6:2 resFTOH.

Our work highlights that degFTOH overtakes resFTOH as the main source of FTOHs. This finding could partially explain the recent rebound of declining atmospheric concentrations of 8:2 FTOH; for instance, a multiannual trend analysis indicated that 8:2 FTOH concentrations in samples from the Global Atmospheric Passive Sampling (GAPS) Network initially declined from 2006 to 2008 but then increased again from 2009 to 2011 [39]. From a regulatory perspective, this finding implies that the increasing contribution of FTP degradation has the potential to partially offset the reduction in FTOH residuals in products achieved by the 2010/15 PFOA Stewardship Program and other regulatory efforts.

4.5   FTP Degradation in Waste Stocks as a Long-Term Source of PFCAs

As indicated above, the use of fluorotelomer-based substances (e.g., AFFFs and surfactants) brings about three types of PFCA emissions, i.e., deg-degPFCAs (Fig. 4.2d and h), as well as impPFCAs and deg-resPFCAs (both liberated from waste stocks and formed in the environment). The releases of the latter two, i.e., impPFOA and deg-resPFOA, are anticipated to cease within a decade when the service life of non-polymeric C8 fluorotelomer-based substances comes to an end. By contrast, the release of deg-degPFOAs will last for decades and even centuries (Fig. 4.2d).

In addition to the use of fluorotelomer-based substances, there exist a number of other PFCA sources [8, 9] (Fig. 4.1). The most recent reliable estimate of the total global release of PFCAs (C4–C14) from these sources is approximately 5600–13000 t between 1951 and 2015 (under the plausible scenario) [9], which is nearly an order of magnitude higher than the estimated cumulative release of deg-degPFCAs (556–591 t, C4–C12) in our highest-release Scenario I for the same period. Therefore, omitting the contribution from the degradation of FTPs when estimating historical PFCA emissions [8, 9, 33] should not lead to a significant underestimation.

Nevertheless, the degradation of FTPs in waste stocks can be a significant source of PFCAs in the future, particularly when deliberate uses of (long-chain) PFCAs will terminate. Our estimates suggest that 1323–1535 t of deg-degPFCAs (C4–C12) will be released during 2016–2040 in Scenario I (data not shown here). This level is in the same order as the projected cumulative release of 10–3830 t from intentional uses of PFCAs (C4–C14) for 2015–2030 (worst case assuming no restrictions on PFCAs will be taken in developing countries) [9]. Meanwhile, our long-term calculation under Scenario I (data not shown here) indicates that, between 2015 and 2100, another seven times as much will be released as the cumulative emissions of deg-degPFOA as of 2015. Moreover, by 2100, 1170–1380 t of deg-degPFOA is estimated to be present in waste stocks, which is available for leaching into the environment in the following centuries. Our calculations highlight the need for environmentally sound management of the waste stocks of FTPs into the foreseeable future. On the one hand, instead of being deposited in landfills and dumps, destruction and irreversible transformation techniques (e.g., incineration) should be the preferable disposal options for obsolete FTP-containing finished products. On the other hand, landfill gases and leachates from historical and current landfills and dumps, which are, respectively, the two major routes liberating degFTOHs and deg-degPFCAs into the environment, should be appropriately treated or remediated.

4.6   Summary

Our modeling quantitatively depicts the entire story that FTP degradation in waste stocks is making an increasing contribution to FTOH generation, the bulk of which readily migrates from waste stocks and is degraded into PFCAs in the environment; the remaining part of the generated FTOHs is degraded in waste stocks, which makes those stocks reservoirs that slowly release PFCAs into the environment over the long run because of the low leaching rate and extreme persistence of PFCAs. The degradation of FTPs leads to up to 130–160 t of 8:2FTOH emissions and 7–9 t/a of PFOA emissions every year, in the scenario where a short product lifespan (LS) and a short degradation half-life (HL) in waste stocks are assumed. When both are at possibly realistic levels, LS is more crucial than HL to an accurate description of FTPs stocks, whereas the emissions of degFTOHs and deg-degPFCAs are more sensitive to HL. While the relative importance of FTOH generation in waste stocks and physical environment in the total FTOH emissions is presently inconclusive because of large uncertainty ranges, PFCA formation in the physical environment is identified as a more relevant PFCA source than leaching from waste stocks. Our preliminary calculations highlight the need for environmentally sound management of obsolete FTP-containing products into the foreseeable future.

References

1. USEPA (2002) Telomer research program update. Presentation to USEPA-OPPT, November 25, 2002; U.S. Public Docket AR226-1141
2. OECD (2013) OECD/UNEP global PFC group, synthesis paper on per- and polyfluorinated chemicals (PFCs). Environment, Health and Safety, Environment Directorate, OECD
3. Washington JW, Jenkins TM, Rankin K, Naile JE (2015) Decades-scale degradation of commercial, side-chain, fluorotelomer-based polymers in soils and water. Environ Sci Technol 49(2):915–923
4. Russell MH, Berti WR, Szostek B, Buck RC (2008) Investigation of the biodegradation potential of a fluoroacrylate polymer product in aerobic soils. Environ Sci Technol 42(3):800–807
5. Washington JW, Ellington JJ, Jenkins TM, Evans JJ, Yoo H, Hafner SC (2009) Degradability of an acrylate-linked, fluorotelomer polymer in soil. Environ Sci Technol 43(17):6617–6623
6. Ellis DA, Martin JW, De Silva AO, Mabury SA, Hurley MD, Sulbaek Andersen MP, Wallington TJ (2004) Degradation of fluorotelomer alcohols: a likely atmospheric source of perfluorinated carboxylic acids. Environ Sci Technol 38(12):3316–3321
7. Wang Z, Cousins IT, Scheringer M, Buck RC, Hungerbühler K (2014) Global emission inventories for C4–C14 perfluoroalkyl carboxylic acid (PFCA) homologues from 1951 to 2030, part II: the remaining pieces of the puzzle. Environ Int 69:166–176
8. Prevedouros K, Cousins IT, Buck RC, Korzeniowski SH (2006) Sources, fate and transport of perfluorocarboxylates. Environ Sci Technol 40(1):32–44
9. Wang Z, Cousins IT, Scheringer M, Buck RC, Hungerbühler K (2014) Global emission inventories for C4–C14 perfluoroalkyl carboxylic acid (PFCA) homologues from 1951 to 2030, part I: production and emissions from quantifiable sources. Environ Int 70:62–75
10. Royer LA, Lee LS, Russell MH, Nies LF, Turco RF (2015) Microbial transformation of 8: 2 fluorotelomer acrylate and methacrylate in aerobic soils. Chemosphere 129:54–61
11. Kim M, Li LY, Grace JR, Yue C (2015) Selecting reliable physicochemical properties of perfluoroalkyl and polyfluoroalkyl substances (PFASs) based on molecular descriptors. Environ Pollut 196:462–472
12. Armitage JM, MacLeod M, Cousins IT (2009) Modeling the global fate and transport of perfluorooctanoic acid (PFOA) and perfluorooctanoate (PFO) emitted from direct sources using a multispecies mass balance model. Environ Sci Technol 43(4):1134–1140
13. Goss K-U (2008) The pK_a values of PFOA and other highly fluorinated carboxylic acids. Environ Sci Technol 42(2):456–458
14. MacLeod M, Scheringer M, Hungerbühler K (2007) Estimating enthalpy of vaporization from vapor pressure using Trouton's rule. Environ Sci Technol 41(8):2827–2832
15. Ellis D, Martin J, Mabury S, Hurley M, Sulbaek Andersen M, Wallington T (2003) Atmospheric lifetime of fluorotelomer alcohols. Environ Sci Technol 37(17):3816–3820
16. Zhao L, McCausland PK, Folsom PW, Wolstenholme BW, Sun H, Wang N, Buck RC (2013) 6: 2 Fluorotelomer alcohol aerobic biotransformation in activated sludge from two domestic wastewater treatment plants. Chemosphere 92(4):464–470
17. Wang N, Szostek B, Folsom PW, Sulecki LM, Capka V, Buck RC, Berti WR, Gannon JT (2005) Aerobic biotransformation of ^{14}C-labeled 8-2 telomer B alcohol by activated sludge from a domestic sewage treatment plant. Environ Sci Technol 39(2):531–538
18. Liu J, Wang N, Szostek B, Buck RC, Panciroli PK, Folsom PW, Sulecki LM, Bellin CA (2010) 6-2 Fluorotelomer alcohol aerobic biodegradation in soil and mixed bacterial culture. Chemosphere 78(4):437–444
19. Wang N, Szostek B, Buck RC, Folsom PW, Sulecki LM, Gannon JT (2009) 8-2 Fluorotelomer alcohol aerobic soil biodegradation: pathways, metabolites, and metabolite yields. Chemosphere 75(8):1089–1096
20. Zhao L, Folsom PW, Wolstenholme BW, Sun H, Wang N, Buck RC (2013) 6:2 Fluorotelomer alcohol biotransformation in an aerobic river sediment system. Chemosphere 90(2):203–209

21. Hurley M, Sulbaek Andersen M, Wallington T, Ellis D, Martin J, Mabury S (2004) Atmospheric chemistry of perfluorinated carboxylic acids: reaction with OH radicals and atmospheric lifetimes. J Phys Chem A 108(4):615–620
22. Vaalgamaa S, Vähätalo AV, Perkola N, Huhtala S (2011) Photochemical reactivity of perfluorooctanoic acid (PFOA) in conditions representing surface water. Sci Total Environ 409(16):3043–3048
23. Armitage J, Cousins IT, Buck RC, Prevedouros K, Russell MH, MacLeod M, Korzeniowski SH (2006) Modeling global-scale fate and transport of perfluorooctanoate emitted from direct sources. Environ Sci Technol 40(22):6969–6975
24. Li L, Liu J, Hu J, Wania F (2017) Degradation of fluorotelomer-based polymers contributes to the global occurrence of fluorotelomer alcohol and perfluoroalkyl carboxylates: a combined dynamic substance flow and environmental fate modeling analysis. Environ Sci Technol 51(8):4461–4470
25. Li L, Zhai Z, Liu J, Hu J (2015) Estimating industrial and domestic environmental releases of perfluorooctanoic acid and its salt in China. Chemosphere 129:100–109
26. Russell MH, Berti WR, Szostek B, Wang N, Buck RC (2010) Evaluation of PFO formation from the biodegradation of a fluorotelomer-based urethane polymer product in aerobic soils. Polym Degrad Stab 95(1):79–85
27. ECHA (2014) ANNEX XV restriction report proposal for a restriction. Perfluorooctanoic Acid (PFOA), PFOA salts and PFOA-related substances. European Chemicals Agency (ECHA), Helsinki, Finland
28. USEPA (2009) Long-chain perfluorinated chemicals (PFCs) action plan. U.S. Environmental Protection Agency, Washington D.C.
29. Castelvetro V, Aglietto M, Ciardelli F, Chiantore O, Lazzari M, Toniolo L (2002) Structure control, coating properties, and durability of fluorinated acrylic-based polymers. J Coating Technol 74(928):57–66
30. DuPont EI (1987) Oil- and water-repellent copolymers (EU Patent No. 87300551.6)
31. Watkins MH, Covelli CA, Seals LR (2006) Fabric treated with durable stain repel and stain release finish and method of industrial laundering to maintain durability of finish (US Patent No. 2006/0228964 A1)
32. van Zelm R, Huijbregts MA, Russell MH, Jager T, van de Meent D (2008) Modeling the environmental fate of perfluorooctanoate and its precursors from global fluorotelomer acrylate polymer use. Environ Toxicol Chem 27(11):2216–2223
33. Wania F (2007) A global mass balance analysis of the source of perfluorocarboxylic acids in the Arctic Ocean. Environ Sci Technol 41(13):4529–4535
34. Wallington T, Hurley M, Xia J, Wuebbles D, Sillman S, Ito A, Penner J, Ellis D, Martin J, Mabury S (2006) Formation of $C_7F_{15}COOH$ (PFOA) and other perfluorocarboxylic acids during the atmospheric oxidation of 8: 2 fluorotelomer alcohol. Environ Sci Technol 40(3):924–930
35. Schenker U, Scheringer M, Macleod M, Martin JW, Cousins IT, Hungerbühler K (2008) Contribution of volatile precursor substances to the flux of perfluorooctanoate to the Arctic. Environ Sci Technol 42(10):3710–3716
36. Yan H, Cousins IT, Zhang C, Zhou Q (2015) Perfluoroalkyl acids in municipal landfill leachates from China: occurrence, fate during leachate treatment and potential impact on groundwater. Sci Total Environ 524–525:23–31
37. Yarwood G, Kemball-Cook S, Keinath M, Waterland RL, Korzeniowski SH, Buck RC, Russell MH, Washburn ST (2007) High-resolution atmospheric modeling of fluorotelomer alcohols and perfluorocarboxylic acids in the North American troposphere. Environ Sci Technol 41(16):5756–5762
38. Wang Z, Xie Z, Mi W, Möller A, Wolschke H, Ebinghaus R (2015) Neutral poly/per-fluoroalkyl substances in air from the Atlantic to the Southern Ocean and in Antarctic snow. Environ Sci Technol 49(13):7770–7775
39. Gawor A, Shunthirasingham C, Hayward S, Lei Y, Gouin T, Mmereki B, Masamba W, Ruepert C, Castillo L, Shoeib M, Lee SC, Harner T, Wania F (2014) Neutral polyfluoroalkyl substances in the global atmosphere. Environ Sci Process Impact 16(3):404–413

Chapter 5
Elucidating the Variability in the Hexabromocyclododecane Diastereomer Profile in the Global Environment

Abstract Hexabromocyclododecane (HBCDD) is a hazardous flame-retardant subject to international regulation. Whereas γ-HBCDD is a dominant component in the technical HBCDD mixture, the diastereomer profile in environmental samples shows substantial temporal (e.g., present vs. future) and spatial (e.g., industrial, populated vs. remote background areas) variations, ranging from γ- to α-HBCDD-dominant. In this chapter, we explain the variability in the diastereomer profile of HBCDD around the world by modeling the global emissions and fate of HBCDD diastereomers. Our modeling results indicate that, as of 2015, 340–1000 tonnes of HBCDD have been emitted globally, with slightly more γ-HBCDD (50–65%) than α-HBCDD (30–50%). Emissions of γ-HBCDD primarily originate from production and other industrial processes ("industrial sources"), whereas those of α-HBCDD are mainly associated with the use and end-of-life disposal of HBCDD-containing products ("consumer sources"). Presently, α-HBCDD dominates the contamination in the air of populated areas, while γ-HBCDD is more abundant in remote background areas and in regions with HCBDD production and processing facilities. Globally, the relative abundance of α-HBCDD is anticipated to increase after production of HBCDD is banned. Given that α-HBCDD is more persistent and bioaccumulative than other diastereomers, isomerization has bearing on the potential environmental and health impacts on a global scale.

5.1 Introduction

The past decades have witnessed extensive use of hexabromocyclododecane (HBCDD) as a brominated flame retardant in construction materials and household products, such as expanded (EPS) and extruded (XPS) polystyrene insulation boards. The worldwide use and emissions of this compound, as well as subsequent long-range environmental transport, result in its ubiquity in a diversity of abiotic and biotic compartments around the world [1–3]. Worldwide concern over HBCDD arose because it is persistent in the environment, bioaccumulative, and toxic to terrestrial and aquatic organisms [4]. The Stockholm Convention has listed HBCDD as a persistent organic pollutant in 2013 for international restrictions on its production, trade and use.

© Springer Nature Singapore Pte Ltd. 2020

L. Li, *Modeling the Fate of Chemicals in Products*, Springer Theses,
https://doi.org/10.1007/978-981-15-0579-9_5

Technical HBCDD is a mixture synthesized via bromination of cyclododeca-1,5,9-trienes and theoretically consists of 16 possible stereoisomers [5]. Among them, γ-HBCDD is the most abundant (with a typical mass fraction of 81.6%), followed by α-HBCDD (11.8%) and β-HBCDD (5.8%) [5]. Such a γ-HBCDD-dominant diastereomer profile coincides with the dominance of γ-HBCDD observed in many abiotic environmental samples collected from *industrial sites* such as those in Weifang (China) [6] and Tianjin (China) [7]. Meanwhile, the γ-HBCDD-dominant diastereomer profile also seems prevalent in *remote background areas* such as Ny-Ålesund (Norway) [8] and Sleeping Bear Dunes (the United States) [9], which are free from industrial activities. More often than not, α-HBCDD is identified as the dominant component in abiotic environmental samples elsewhere, in particular, *populated or urban areas*. For example, α-HBCDD accounts for 59–68% of total HBCDD in the atmosphere in Guangzhou (China) [10] and on average 49% (12–85%) in soils in Shanghai (China) [11]. In addition to the divergent diastereomer profiles, temporal trends of contamination are also inconsistent between diastereomers. For instance, during the past decade, γ-HBCDD concentration in lake trout from Lake Ontario declined at an annual rate of ~20%, which is three times faster than that of α-HBCDD (6%) [12]. In other words, α-HBCDD is becoming relatively more abundant than γ-HBCDD. It is therefore of interest to establish what factors lead to the spatial and temporal variations in the relative abundance of HBCDD diastereomers in the environment. Obviously, such variations cannot be simply explained by stereo-isometric structural differences, because diastereomers differ marginally in physicochemical properties (e.g., partition tendency and degradability) and hence environmental fate (e.g., long-range transport potential) [13].

Recent lab experiments have confirmed interconversion between diastereomers upon exposure to heat and radiation [14–17]. The transformation from γ-HBCDD to α-HBCDD, i.e., isomerization, is predominant, resulting in an enrichment of α-HBCDD in products [18, 19]. For instance, Ni et al. [20] observed 12 times higher emission of α-HBCDD than γ-HBCDD during open burning of γ-HBCDD-rich polystyrene plastics. As such, the production and use of heat-treated HBCDD-containing products, instead of the technical mixture itself, can be hypothesized to explain the dominance of α-HBCDD observed in the environment. To date, it remains unknown to what extent the isomerization influences the fate of HBCDD diastereomers worldwide, and in particular, how such small-scale processes interact with environmental factors to shape the big picture of HBCDD emissions and occurrence on a global scale. Motivated by these questions, in this chapter, we will illustrate how the coupled CiP-CAFE and BETR-Global modeling helps address the substantial spatial (e.g., industrial, populated vs. remote background areas) and temporal (e.g., present vs. future) variations in the diastereomer profile of HBCDD around the world.

5.2 Method and Data

5.2.1 Properties of HBCDD Diastereomers

In this work, we focus on three major HBCDD diastereomers: α-HBCDD (CAS no. 134237-50-6), β-HBCDD (CAS no. 134237-51-7) and γ-HBCDD (CAS no. 134237-52-8). The total concentration of the three diastereomers is collectively referred to as Σ_3HBCDD hereafter. For each congener, the partition coefficients K_{AW} and K_{OA} at 25 °C are recommended values from Arnot et al. [13], which have been adjusted for thermodynamic consistency. Internal energies of phase transfer, ΔU_W, ΔU_A, and ΔU_O, which are used to adjust the partition coefficients for temperature dependence, are calculated in accordance with MacLeod et al. [21] based on Trouton's Rule and defaults. Degradation half-lives in soil (and solid waste in landfill, and dumping and simple landfill), water (and wastewater) and air at 25 °C are taken from Arnot et al. [13]. Activation energies for adjusting the degradation half-lives for temperature dependence are defaults in the CiP-CAFE and BETR-Global models.

5.2.2 Production, Applications and Interregional Trade

Production of the technical HBCDD mixture in individual CiP-CAFE regions as of 2015 was gathered from different sources (for details, see Li et al. [22]). The compiled information indicates that cumulatively 40% of production occurred in North America, followed by 27% in China and 25% in Western Europe. Given that China is likely currently the only producer and the Stockholm Convention tentatively allows a five-year "specific exemption" for China with ongoing production and exclusive use in building insulation, we assume that production in China will linearly decrease after 2016 and cease in 2021. This assumption leads to a cumulative production of 713 kilotonnes (kt) of the technical HBCDD mixture worldwide for the entire modeled period. The production of three diastereomers is then estimated by multiplying weight fractions in the technical HBCDD mixture (i.e., 11.8% α-HBCDD, 5.8% β-HBCDD, and 81.6% γ-HBCDD [5]) with this total, yielding estimates of 84 kt α-HBCDD, 41 kt β-HBCDD and 582 kt γ-HBCDD. Given that the diastereomer composition was similar in technical HBCDD mixtures purchased from major producers worldwide (9.5–13.3% α-HBCDD, 0.5–11.7% β-HBCDD, and 72.3–88.5% γ-HBCDD) [16], we do not consider its variation between production regions and over time. Note that the sum of the diastereomer-specific production amounts is not exactly equal to the production of the technical HBCDD mixture due to omission of minor stereoisomers, e.g., 0.5% δ-HBCDD and 0.3% ε-HBCDD. Figure 5.1 presents the temporal evolution of the global production of the three major HBCDD diastereomers from six regions.

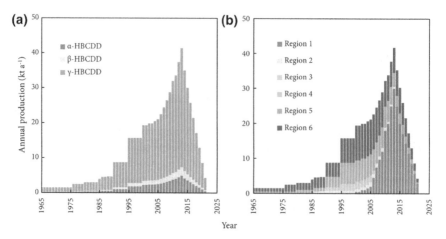

Fig. 5.1 The production history of three HBCDD diastereomers (panel **a**) from six regions (panel **b**) worldwide. Adapted and reproduced with permission (Li and Wania [22]) Copyright (2018) American Chemical Society

   While HBCDD appears in a vast diversity of products, it is predominantly used in five applications (APs): EPS-insulation boards (AP1) and XPS-insulation boards (AP2) in buildings, textile coating agent (AP3), EPS-package (AP4), high impact polystyrene (HIPS) in electrical and electronic equipment (AP5). The relative distribution of HBCDD consumption among the applications in individual regions is also compiled from the literature (for details, see Li et al. [22]). The compiled data indicate that AP1 and AP2 are dominant applications in all regions, e.g., accounting for >98% of the total HBCDD consumption in Region 1 (mainland China), ~85% in Region 2 (Asia-Pacific) and >90% in Region 5 (Western Europe). Based on those data, CiP-CAFE calculates that insulation boards in buildings (AP1 and AP2) account for >97% of global HBCDD consumption.

   CiP-CAFE also requires information on the lifespan of products in each application. For AP1 and AP2 where HBCDD is embedded in insulation boards in buildings, we use region-specific building lifespan distributions (Fig. 5.2), ranging from 23 years in mainland China (Region 1) [23] to ~100 years in Western Europe (Region 5) [24, 25]. We assume constant lifespans throughout the modeled period due to a lack of temporally resolved information. For the remaining APs, we assign the same lifespans to all regions: a Weibull distribution with a shape parameter of 10 years and a shape parameter of 2.4 for textile products in AP3 [26], <1 year for packaging materials in AP4 [27], and a Weibull distribution with a shape parameter of 10 years and a shape parameter of 2.5 for products in AP5 [28].

   We consider the interregional mass exchange of HBCDD through trade flows of (i) the technical HBCDD mixture (after LC1), and (ii) polystyrene beads and polyester textiles (after LC2; data taken from statistical databases [29–32]), and (iii) waste electrical and electronic equipment (after LC5; data taken from Breivik et al. [33]). Due to a lack of temporally resolved data, import and export fractions of (ii) and (iii)

Fig. 5.2 Lifespan distributions (central and deviation parameters in parentheses) of buildings in seven modeled regions (REs) (for details, see Li et al. [22]). Reproduced with permission (Li and Wania [22]) Copyright (2018) American Chemical Society

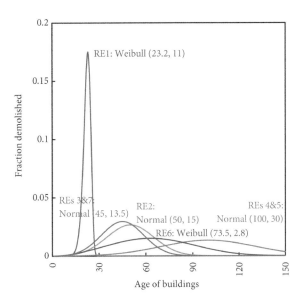

are assumed to be constant throughout the modeled period. We do not consider the international trade of insulation boards in APs 1 and 2 and their associated waste because insulation boards are locally manufactured and rarely traded internationally due to the cost of transporting such bulky commodities [34].

5.2.3   Emission and Waste Factors

Emission and waste factors, most of which are diastereomer-specific, are compiled from the literature [20, 35–37] or computed using built-in modules in CiP-CAFE (Sect. 2.2.3.2). For details, see Li and Wania [22].

The emission factors can be quite variable and uncertain. For example, the emission factors for production (LC1) and industrial processes (LC2) are mainly based on manufacturing conditions in developed countries [35], which may underestimate actual emissions in developing countries that lack proper emission abatement technologies. Meanwhile, when calculating emission factors for in-use insulation boards, the built-in EmissionRate module assumes that HBCDD is released from bare boards. However, in reality, some boards may be covered by foil film, oriented strand boards or mortar. To highlight the uncertainty associated with these emission factors, we defined two additional sets of model scenarios in which the default emission factors (i) for industrial processes are increased by an order of magnitude, and (ii) for in-use stocks of insulation boards in AP1 and AP2 are reduced by a factor of two.

5.2.4 Disposal of HBCDD-Containing Solid Waste

Recent evidence reveals (i) the recycling and reuse (WD5) of waste polystyrene insulation boards (AP1 and AP2) and packaging materials (AP4) [36, 38], which is believed to be the reason for the occurrence of HBCDD in new consumer products that do not need to be flame retarded [39, 40], and (ii) illegal open burning (WD6) of e-waste that leads to substantial emissions of HBCDD [41]. Therefore, in our calculation, we gather information on the fractions of waste from AP1, AP2 and AP4 being recycled and reused (for details, see Li et al. [22]), as well as the fraction of e-waste from AP5 burned in developing regions (Regions 1, 3 and 7) (taken from Breivik et al. [42]), i.e., the waste disposal ratios for WD5 and WD6. We further assume the absence of environmentally sound treatment of HBCDD-containing waste (WD7). The rest of the waste from these applications, and all textile waste from AP3, are assumed to not be distinguishable from general solid waste and thus are distributed among WDs 1, 3 and 4, according to the built-in time-dependent waste disposal ratios for individual regions in CiP-CAFE.

5.2.5 Quantification of Isomerization

The focus of this chapter lies on the fate of, and the interconversion between, three HBCDD diastereomers, which, therefore, requires modification of original CiP-CAFE model to accommodate isomerization of HBCDD diastereomers.

5.2.5.1 Isomerization in the Anthroposphere

HBCDD diastereomers isomerize during thermal treatment of HBCDD-containing raw polymeric materials in the "industrial processes (LC2)" stage, resulting in a change in the diastereomer profile, i.e., enrichment of α-HBCDD and a decrease in γ-HBCDD. We can treat this isomerization process as the decomposition of a mother diastereomer and simultaneously the generation of a daughter diastereomer. In other words, whereas the total mass of HBCDD is preserved, the mass of every single diastereomer is not. To characterize the mass change during isomerization from a mother diastereomer to a daughter diastereomer, we define an "isomerization factor (IF)" for each "parent-daughter" pair (Eq. 5.1):

$$\text{IF}_{\text{mother} \to \text{daughter}} = \frac{m_{\text{daughter}}}{m_{\text{mother}}} \tag{5.1}$$

whereby, m_{mother} is the mass of a mother diastereomer (before isomerization), and m_{daughter} is the mass of a daughter diastereomer (after isomerization). For a given mother diastereomer, the IF values of all its daughter diastereomers sum up to 1.

Therefore, the masses of total diastereomers entering the processing stage ($\mathbf{M}_{\text{before}}$), and departing the processing stage and then entering the use phase ($\mathbf{M}_{\text{after}}$), will be linked by an isomerization factor matrix ($\mathbf{IF}$) (Eq. 5.2):

$$\mathbf{M}_{\text{after}} = \mathbf{IF} \cdot \mathbf{M}_{\text{before}} \tag{5.2}$$

or,

$$
\begin{bmatrix} M_{\alpha,\text{after}} \\ M_{\beta,\text{after}} \\ M_{\gamma,\text{after}} \end{bmatrix}
=
\begin{bmatrix}
\text{IF}_{\alpha \to \alpha} & \text{IF}_{\beta \to \alpha} & \text{IF}_{\gamma \to \alpha} \\
\text{IF}_{\alpha \to \beta} & \text{IF}_{\beta \to \beta} & \text{IF}_{\gamma \to \beta} \\
\text{IF}_{\alpha \to \gamma} & \text{IF}_{\beta \to \gamma} & \text{IF}_{\gamma \to \gamma}
\end{bmatrix}
\begin{bmatrix} M_{\alpha,\text{before}} \\ M_{\beta,\text{before}} \\ M_{\gamma,\text{before}} \end{bmatrix}
\tag{5.3}
$$

Mathematically, $\mathbf{IF}$ denotes a transformation matrix that converts a diastereomer profile to its corresponding isomerized form under specific conditions (e.g., thermal treatment at a given temperature for a given period of time), with the mass balance of the sum of all diastereomers being preserved. We modify the original algorithm in CiP-CAFE to accommodate Eq. (5.3) in the model calculation, with $\mathbf{IF}$ as additional model inputs.

The next question is how to obtain $\mathbf{IF}$ for products in each application. Here we seek to link $\mathbf{IF}$ with the "typical" diastereomer profile observed in products in an application. Let either side of Eq. (5.3) be normalized by the total mass of three diastereomers (M), yielding

$$
\begin{bmatrix} \frac{M_{\alpha,\text{after}}}{M} \\ \frac{M_{\beta,\text{after}}}{M} \\ \frac{M_{\gamma,\text{after}}}{M} \end{bmatrix}
=
\begin{bmatrix}
\text{IF}_{\alpha \to \alpha} & \text{IF}_{\beta \to \alpha} & \text{IF}_{\gamma \to \alpha} \\
\text{IF}_{\alpha \to \beta} & \text{IF}_{\beta \to \beta} & \text{IF}_{\gamma \to \beta} \\
\text{IF}_{\alpha \to \gamma} & \text{IF}_{\beta \to \gamma} & \text{IF}_{\gamma \to \gamma}
\end{bmatrix}
\begin{bmatrix} \frac{M_{\alpha,\text{before}}}{M} \\ \frac{M_{\beta,\text{before}}}{M} \\ \frac{M_{\gamma,\text{before}}}{M} \end{bmatrix}
\tag{5.4}
$$

whereby

$$M = M_{\alpha,\text{after}} + M_{\beta,\text{after}} + M_{\gamma,\text{after}} = M_{\alpha,\text{before}} + M_{\beta,\text{before}} + M_{\gamma,\text{before}} \tag{5.5}$$

or written as:

$$\mathbf{m}_{\text{after}} = \mathbf{IF} \cdot \mathbf{m}_{\text{before}} \tag{5.6}$$

Therefore, we can obtain $\mathbf{IF}$ from the diastereomer profiles before ($\mathbf{m}_{\text{before}}$, i.e., the one in the technical mixture) and after a thermal process ($\mathbf{m}_{\text{after}}$, i.e., the one observed in end products). In other words, we link the macroscopic $\mathbf{M}_{\text{before}}$ and $\mathbf{M}_{\text{after}}$ that lump together the total mass of a diastereomer in the entire anthroposphere, with the microscopic $\mathbf{m}_{\text{before}}$ and $\mathbf{m}_{\text{after}}$ that can be experimentally measured in a wet lab. The IF values (individual elements in the isomerization factor matrix $\mathbf{IF}$) vary between applications because products in different applications are subject to different thermal treatments. If a product is processed at a high temperature over a sufficient length of time (e.g., 2.2 h at 160 °C [19]; shorter if temperature increases), isomerization is under thermodynamic control, resulting in a constant equilibrium diastereomer profile consisting of ~80% α-HBCDD, ~11% β-HBCDD, and ~8% γ-HBCDD [16, 19]. However, most products are processed at a lower temperature (e.g., 80–120 °C) and for a limited period of time, during which isomerization is under kinetic control [14]. This leads to partial isomerization; that is, diastereomer profiles can fall between a "γ-HBCDD-rich" and an "α-HBCDD-rich" profile. Measured profiles with maximum/minimum α-HBCDD and γ-HBCDD mass fractions in EPS (i.e., applicable to APs 1 and 4) [18], XPS (AP2) [18], and textiles (AP3) have been reported in the literature [43]. Based on these reports, we calculate two **IF** matrices, one yielding either an α-HBCDD-rich or a γ-HBCDD-rich scenario, for each of APs 1–4 (Table 5.1). No measurements are currently available for HIPS in AP5; thus, we assume that, in both the α-HBCDD-rich and γ-HBCDD-rich scenarios, products in AP5 always show the equilibrium HBCDD diastereomer profile because HIPS is often manufactured at a high temperature ($\geq$180 °C) (Table 5.1).

We assume no isomerization during the remaining lifecycle stages due to a lack of experimental evidence for it, although we recognize the theoretical possibility of

Table 5.1 Isomerization factor matrixes **IF** (for details, see Li and Wania [22])

		IF (resulting in a γ-rich profile)	IF (resulting in an α-rich profile)
AP1		$\begin{bmatrix} 1.00 & 0 & 0 \\ 0 & 1.00 & 0 \\ 0 & 0 & 1.00 \end{bmatrix}$	$\begin{bmatrix} 0.89 & 0 & 0.43 \\ 0.06 & 0.97 & 0.32 \\ 0.05 & 0.03 & 0.25 \end{bmatrix}$
AP2		$\begin{bmatrix} 0.88 & 0 & 0.39 \\ 0.06 & 0.97 & 0.36 \\ 0.06 & 0.03 & 0.25 \end{bmatrix}$	$\begin{bmatrix} 0.88 & 0 & 0.39 \\ 0.06 & 0.97 & 0.36 \\ 0.06 & 0.03 & 0.25 \end{bmatrix}$
AP3		$\begin{bmatrix} 0.90 & 0.18 & 0.18 \\ 0.04 & 0.78 & 0.10 \\ 0.06 & 0.04 & 0.72 \end{bmatrix}$	$\begin{bmatrix} 0.90 & 0.14 & 0.42 \\ 0.04 & 0.82 & 0.13 \\ 0.06 & 0.04 & 0.45 \end{bmatrix}$
AP4		$\begin{bmatrix} 1.00 & 0 & 0 \\ 0 & 1.00 & 0 \\ 0 & 0 & 1.00 \end{bmatrix}$	$\begin{bmatrix} 0.89 & 0 & 0.43 \\ 0.06 & 0.97 & 0.32 \\ 0.05 & 0.03 & 0.25 \end{bmatrix}$
AP5		$\begin{bmatrix} 0.90 & 0.19 & 0.80 \\ 0.04 & 0.77 & 0.11 \\ 0.06 & 0.04 & 0.09 \end{bmatrix}$	$\begin{bmatrix} 0.90 & 0.19 & 0.80 \\ 0.04 & 0.77 & 0.11 \\ 0.06 & 0.04 & 0.09 \end{bmatrix}$
Environment		$\begin{bmatrix} 0.90 & 0.24 & 0.32 \\ 0.04 & 0.71 & 0.06 \\ 0.06 & 0.05 & 0.62 \end{bmatrix}$	$\begin{bmatrix} 0.90 & 0.24 & 0.32 \\ 0.04 & 0.71 & 0.06 \\ 0.06 & 0.05 & 0.62 \end{bmatrix}$

photolytically mediated isomerization. For example, HBCDD in textiles is not found to isomerize during a 371-day exposure to sunlight, possibly due to a protective effect by the polymeric fibers in textiles [37]. In addition, we do not consider isomerization during open burning because the isomerization-induced enrichment of α-HBCDD is already reflected in the emission factors in the literature [20].

5.2.5.2   Isomerization in the Physical Environment

Earlier monitoring work has reported photo-isomerization of HBCDD diastereomers in the environment [15], which is considered in our BETR-Global simulation. It is observed that photo-isomerization is rapid and reaches a quasi-equilibrium diastereomer profile within 4–7 days [15], which is shorter than the average atmospheric residence time (7.5 days) in a single BETR-Global grid cell [44]. That is, statistically speaking, photo-isomerization is largely complete before HBCDD departs from the grid cell into which it is emitted. Therefore, for each grid, we convert the original CiP-CAFE-derived emission estimate ($E_{CiP\text{-}CAFE}$) to an isomerized emission estimate, which serves as the input to BETR-Global (E_{BETR}):

$$E_{BETR} = IF \cdot E_{CiP\text{-}CAFE} \qquad (5.7)$$

That is,

$$\begin{bmatrix} E_{\alpha,BETR} \\ E_{\beta,BETR} \\ E_{\gamma,BETR} \end{bmatrix} = \begin{bmatrix} IF_{\alpha\to\alpha} & IF_{\beta\to\alpha} & IF_{\gamma\to\alpha} \\ IF_{\alpha\to\beta} & IF_{\beta\to\beta} & IF_{\gamma\to\beta} \\ IF_{\alpha\to\gamma} & IF_{\beta\to\gamma} & IF_{\gamma\to\gamma} \end{bmatrix} \cdot \begin{bmatrix} E_{\alpha,CiP\text{-}CAFE} \\ E_{\beta,CiP\text{-}CAFE} \\ E_{\gamma,CiP\text{-}CAFE} \end{bmatrix} \qquad (5.8)$$

The isomerization factor matrix **IF** in Eq. (5.7) is defined in the same manner as that in Eq. (5.2). We calculate the **IF** for photo-isomerization in the environment based on experimental data in Harrad et al. [15] (Table 5.1).

In our current work, we neglect biologically mediated isomerization, e.g., by microorganisms in sewage sludge and the natural rhizosphere, because its environmental relevance is still uncertain and mechanistic information required for modeling is inadequate [4, 45]. For example, isomerization from γ-HBCDD to α-HBCDD by sludge microbes was found to be statistically insignificant by Davis et al. [46] but to lead to a nearly 2-fold increase in abundance of α-HBCDD by Gerecke et al. [47].

5.2.6   Uncertainty Scenarios

As elaborated above, we consider the uncertainties/variations in industrial emission factors (2 scenarios), emission factors from in-use stock (2 scenarios), and IF (2 scenarios). Their combination results in eight model scenarios, which are run separately in CiP-CAFE and BETR-Global. In the following, each estimate is presented as a range, with the maximum of the eight scenarios designated as an upper bound and the minimum designated as a lower bound.

5.3   Evaluation of Modeling Performance

To evaluate the performance of our models, we compare modeled atmospheric concentrations of Σ_3HBCDD in BETR-Global grid cells with observations from two sampling campaigns (March 2005 to March 2006 [48], and January 2014 to December 2014 [49]) of the Global Atmospheric Passive Sampling (GAPS) Network (Fig. 5.3). Here, Σ_3HBCDD is used for comparison because diastereomer-specific observations are not available on a global scale. In Fig. 5.3, our modeled concentration ranges overlap with the observed concentration ranges for 32 of the 40 GAPS grid cells in the 2005–2006 dataset, and 26 of the 36 GAPS grid cells in the 2014 dataset. Sixteen medians in the 2005–2006 dataset and 14 medians in the 2014 dataset are within the modeled concentration ranges. Encouragingly, the models succeed in reproducing the observed contamination in several hotspot regions such as Europe (Cells 38, and 61 to 63), North America (Cells 52 to 54, and 76 to 78), and mainland China (Cells 114 and 115), in particular for the 2005–2006 dataset. Modeled concentrations in 2014 (Fig. 5.3b) are around an order of magnitude lower than those in 2005–2006 (Fig. 5.3a), which is in line with the observed decline between the two datasets. As such, the combination of CiP-CAFE and BETR-Global is capable of capturing the temporal and spatial trends in the global HBCDD contamination.

At first glance, the agreement between modeled and observed concentrations is better in the Northern hemisphere (grid cell number ≤ 144) than in the South (grid cell number ≥ 145). In fact, observations in the Southern hemisphere may not be sufficiently reliable because many of them are below the limits of quantification (LOQs) and even below 1/3 of LOQs [49]. Meanwhile, since BETR-Global assumes a homogenous distribution of chemical mass within a grid cell and computes cell-average concentrations, it cannot reproduce extremes observed near local sources. For instance, the surprisingly high concentration in Cell 200 (Fig. 5.3b) was measured in Concepción, i.e., one of the largest conurbations in Chile [49]. In addition, information regarding local sources of HBCDD in the Southern hemisphere is incomplete, which can also contribute to the observed contamination. For example, use of HBCDD-containing insulation materials in research stations is a leading source responsible for HBCDD contamination in Antarctica [50], whereas CiP-CAFE does not take into account Antarctica as a modeled region in its model configuration.

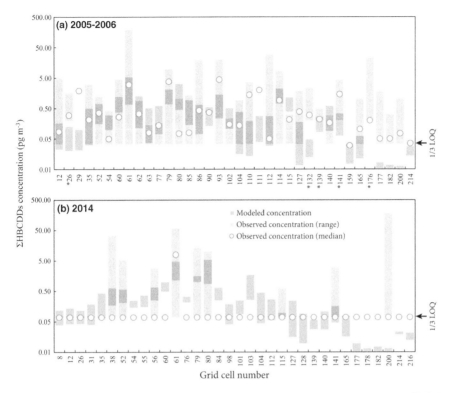

Fig. 5.3 Comparison between modeled atmospheric Σ_3HBCDD concentrations in BETR-Global cells (yellow bars) and observations (blue bars and dots) from the Global Atmospheric Passive Sampling (GAPS) Network in 2005–2006 (panel **a**) and 2014 (panel **b**). Note that the two sampling campaigns did not sample the same set of grid cells. For the numbering of grid cells, see Chap. 2. Observations below limits of quantification (LOQ, 0.1 pg m^{-3} for 2005–2006 samples except for 0.6 pg m^{-3} for grids marked by an asterisk, and 0.21 pg m^{-3} for 2014 samples) are represented by 1/3 of the corresponding LOQs. In each cell, the range of modeled concentration characterizes the temporal variability in ΣHBCDD concentration within the sampling period, whereas the range of observed concentration represents the maximum and minimum of multiple samples within a cell. Reproduced with permission (Li and Wania [22]) Copyright (2018) American Chemical Society

5.4 Global Emissions of HBCDD Diastereomers

With confidence in the model's performance established, we first explore the diastereomeric composition of HBCDD in the anthroposphere and in the global emissions. Our calculation shows that, as of 2015, cumulatively 160–280 kt of α-HBCDD, 115–210 kt of β-HBCDD and 130–345kt of γ-HBCDD depart from production and industrial processes, i.e., enter the use phase, waste or the environment. Compared with the cumulative production of the three diastereomers (Fig. 5.1), these numbers reflect a considerable net generation of α-HBCDD (around 75–200 kt) and β-HBCDD (around 75–170 kt) in the anthroposphere.

Figure 5.4 shows the CiP-CAFE-derived global emissions of HBCDD diastere-omers for the period 1965–2100. While the production and new use of HBCDD are projected to cease in 2021 (Fig. 5.1), emissions can continue for decades and even a century as a result of in-use and waste stocks. We estimate that, as of 2015, the global cumulative HBCDD emissions amount to 340–1000 t (Fig. 5.4), accounting for less than 0.2% of the total cumulative production worldwide as shown in Fig. 5.1. The emissions comprise 170–280 t of α-HBCDD, 30–80 t of β-HBCDD and 140–650 t of γ-HBCDD. When taking the medians of those estimates, we can find that the share of γ-HBCDD (50–65%) in the cumulative global emissions is slightly higher than that of α-HBCDD (30–50%). In addition, during the entire modeled period 1965–2100, the contribution of α-HBCDD (37–57%) to the cumulative emissions can even outweigh that of γ-HBCDD (30–50%). Given that α-HBCDD accounts for merely 11.8% of the technical HBCDD mixture [5], the elevated share of α-HBCDD among the emissions highlights, again, the importance of isomerization in the global anthroposphere.

Whereas the timing of the peak in emissions of individual diastereomers differs between scenarios, they all peak between 2010 and 2020 (Fig. 5.4) as a result of the decline in global production and new use since 2013. Emission of γ-HBCDD is anticipated to decline sharply in response to the decline in production, whereas that of α-HBCDD will decrease slowly and ultimately dominate the emission in the future (Fig. 5.4). To aid the interpretation of the temporal course of diastereomer-specific emissions, Fig. 5.5 presents a source breakdown for each diastereomer. For the entire modeled period 1965–2100, 55–70% of cumulative emission of γ-HBCDD emanate from production (LC1) and industrial processes (LC2), and disposal of

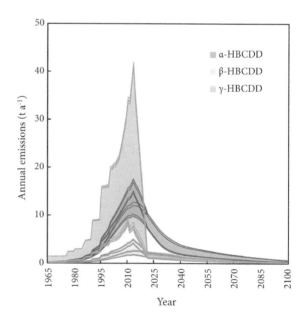

Fig. 5.4 Calculated global emissions of HBCDD diastereomers from 1965 to 2100. For each diastereomer, the colored estimate range is defined by the maximum and minimum of estimates under eight model scenarios. Reproduced with permission (Li and Wania [22]) Copyright (2018) American Chemical Society

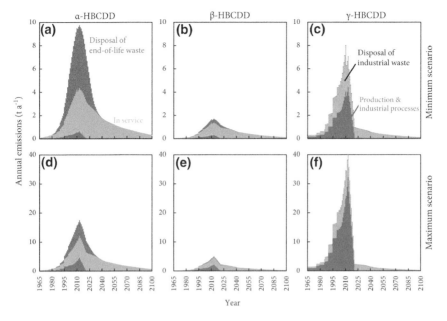

Fig. 5.5 Source-specific global emissions of α-HBCDD (panels **a** and **d**), β-HBCDD (panels **b** and **e**) and γ-HBCDD (panels **c** and **f**) under the minimum (panels **a** to **c**) and maximum (panels **d** to **f**) scenarios. Reproduced with permission (Li and Wania [22]) Copyright (2018) American Chemical Society

industrial waste (Fig. 5.5c and f), while 91–96% of that of α-HBCDD (Fig. 5.5a and d) and 64–80% of those of β-HBCDD (Fig. 5.5b and e) are associated with the use (LC4) and disposal of end-of-life HBCDD-containing products. In other words, the dominance of γ-HBCDD is an indicator of "industrial" sources, whereas that of α-HBCDD reflects "consumer" sources. As shown in Fig. 5.5, although emissions from industrial sources have declined and are expected to end in the near future, those from in-use and waste stocks can continue for decades and even centuries. That is, in terms of the global annual emissions, we can anticipate a transition from the dominance of "industrial" sources to that of "consumer" sources, which explains the switch of the leading role from γ-HBCDD to α-HBCDD as shown in Fig. 5.4. Such a relationship between emission source and dominant diastereomer agrees with observations in the vicinity of sources. For example, Zhang et al. [6] observed the dominance of γ-HBCDD in soils surrounding HBCDD production facilities in Weifang (on average 71.8%) in China, whereas Gao et al. [41] found that α-HBCDD is "obviously abundant" in soil samples collected from illegal e-waste treatment sites.

In terms of Σ_3HBCDD, for the entire modeled period 1965–2100, cumulatively 290–620 t of emissions occur during the use of HBCDD-containing products (orange shades in Fig. 5.5), and 145–150 t of emissions arise from disposal of these products at the end of life (red shades in Fig. 5.5). In addition, a closer inspection of contributions of individual products (data not shown) demonstrates that EPS and XPS insulation boards contribute >95% to emissions from the use phase (in service) because >97% of HBCDD is used in these two applications (Sect. 5.2.2). However, since these estimates are associated with wide uncertainty ranges, it is difficult to conclude whether the cumulative emissions from the two "consumer" sources are larger than those from "industrial" sources, such as production and industrial activities (95–630 t; blue shades in Fig. 5.5) and disposal of industrial waste (60–190 t; light blue shades in Fig. 5.5). It is also noteworthy that the relative importance of industrial and consumer sources differs between regions. For example, our models predict that consumer sources (280 t; range 110–450 t) contribute seven times more to the cumulative emissions in Region 6 (North America) than industrial sources (40 t; range 10–70 t), whereas consumer (140 t; range 110–170 t) and industrial sources (100 t; range 10–200 t) contribute to a similar extent to cumulative emissions in Region 1 (mainland China). The more notable consumer source contribution in North America is mainly due to the three times longer building lifespan in this region (Fig. 5.2) because a longer lifespan delays the HBCDD mass flow entering waste stock and thus amplifies the size of in-use stock.

5.5   Occurrence of HBCDD Diastereomers in the Global Environment

Fed with global emission estimates, BETR-Global simulates the time-dependent atmospheric concentrations of three diastereomers in individual grid cells. We limit the following discussion to the relative abundance of α-HBCDD and γ-HBCDD, as these two diastereomers are predominantly involved in isomerization whereas β-HBCDD is of less importance. Figure 5.6 visualizes the geographic variation of the concentration ratio between α-HBCDD and γ-HBCDD in the lower atmosphere; an α/γ ratio greater than 1 indicates the dominance of α-HBCDD. The prevalence of orange color (indicative of an α/γ ratio between 0.3 and 1) in Fig. 5.6a indicates that γ-HBCDD is more abundant than α-HBCDD, but not much so, in most cells, in particular in remote background areas without significant emissions such as the Arctic and above low- to mid-latitude oceans. The dominance of γ-HBCDD in remote background areas mirrors the slightly higher abundance of γ-HBCDD over α-HBCDD in the cumulative *global* emissions up to 2015 (Fig. 5.3). Our predicted α/γ ratio range aligns with earlier observations in remote background areas, e.g., Ny-Ålesund in Arctic Norway (0.3) [8] and Sleeping Bear Dunes in the United States (0.8) [9]. In addition, the dominance of γ-HBCDD is also observed in cells with significant HBCDD production and processing, e.g., Cell 92 (white box in Fig. 5.6a) encompassing Northern China which is the hub of Chinese HCBDD production [6]. By contrast, α-HBCDD is dominant in the atmosphere of several populated areas such as East and Southeast Asia, Western Europe and North America, because these areas are strongly influenced by emissions from the use and disposal of HBCDD-containing products, which are dominated by α-HBCDD. For example, our model predicts an α/γ ratio of 3.5 for Cell 116 (white box in Fig. 5.6a), which lies within the range of 2.2–3.6 observed in urban and suburban sites of Guangzhou (China) [10]. In fact, the dominance of α-HBCDD is also apparent in a number of indoor dust samples collected from, e.g., Stockholm (Sweden) [51] and Antwerp (Belgium) [52]. This result also explains an earlier finding by Roosens et al. [52] that α-HBCDD is the solely detected diastereomer in the serum of urban residents, and its concentration is significantly correlated with human uptake via indoor dust (dominated by α-HBCDD) rather than dietary items originating from rural areas (dominated by γ-HBCDD).

In 2030 when global HBCDD production is expected to have ended, α-HBCDD will be the dominant diastereomer across the global atmosphere (Fig. 5.6b), because of the transition from "industrial" to "consumer" emissions occurring prior to 2030. Compared with the 2015 case, the α/γ ratio increases in almost all grid cells in 2030; the increase is more prominent in the Northern hemisphere than in the South because

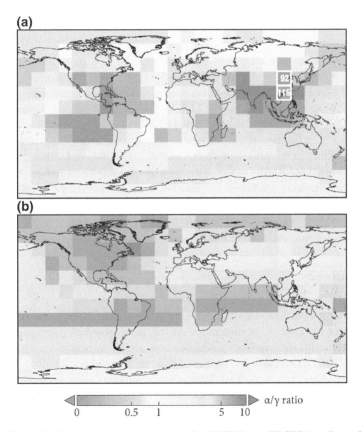

Fig. 5.6 The ratio of atmospheric concentrations of α-HBCDD to γ-HBCDD (medians of concentrations under the eight scenarios) in 288 BETR-Global grid cells in 2015 (panel **a**) and 2030 (panel **b**). White boxes indicate Cells 92 and 116 mentioned in the chapter. Reproduced with permission (Li and Wania [22]) Copyright (2018) American Chemical Society

emissions of α-HBCDD primarily occur in the North. At that moment, the α/γ ratio in the Arctic will still reflect the relative importance of the two diastereomers in the cumulative global emissions, which is predicted to exceed 1. The α/γ ratio is higher in North America than in other populated regions such as East and Southeast Asia. In a word, our modeling results highlight the ever-growing importance of α-HBCDD in future contamination.

Our work reveals that (i) the isomerization-induced generation of α-HBCDD in the anthroposphere and in the global environment is important, and (ii) as a result α-HBCDD will become the dominant HBCDD component when production is terminated. In other words, whereas γ-HBCDD dominates the technical mixture, α-HBCDD can be prominent in the environment, in particular, in remote ecosystems and in the future. This finding intensifies the environmental concern about HBCDD, given that existing evidence suggests that α-HBCDD may be more problematic than γ-HBCDD. For instance, an in vitro assay finds that α-HBCDD is resistant to biotransformation by liver microsomes from marine mammals, while β-HBCDD and γ-HBCDD are quickly metabolized within 90 min [53]. In zebrafish, α-HBCDD is three times more bioaccumulative (an average biomagnification factor of 29.7 vs. 7.61) and persistent (an average half-life of 17.3 vs. 6.08 days) than γ-HBCDD when the concentration in dosed food is low [54]. As such, the isomerization and enrichment of α-HBCDD in the anthroposphere and environment elevate the risk of exposure of biota and humans to HBCDD. Moreover, substantial bio-isomerization of γ-HBCDD, e.g., 96.2–98.6% of γ-HBCDD absorbed by mirror carp converts to α-HBCDD within 60 days [55], can further increase the abundance of α-HBCDD in biota, resulting in the prevalent dominance of α-HBCDD in biota samples from around the world [1, 2, 4]. In this sense, the environmental and health risks of HBCDD can be underestimated if different diastereomers are evaluated as a whole or in the form of the technical mixture. Environmental risk assessment of HBCDD is therefore recommended to be conducted on a diastereomer-specific basis.

5.6 Summary

In this chapter, we elucidate that the dominant release source differs between HBCDD diastereomers because of isomerization in the anthroposphere. Emissions of γ-HBCDD primarily originate from production and other industrial processes, whereas those of α-HBCDD are mainly associated with the use and end-of-life disposal of HBCDD-containing products. This difference in dominant source explains an earlier finding that γ-HBCDD dominates in industrial regions whereas α-HBCDD dominates in populated regions. On a global scale, the cumulative historical emission of γ-HBCDD slightly outweighs that of α-HBCDD, resulting in the dominance of γ-HBCDD in remote background areas. Globally, the relative abundance of α-HBCDD is anticipated to increase after production of HBCDD is banned. Given that α-HBCDD is more persistent and bioaccumulative than other diastereomers, our work suggests that isomerization has bearing on the potential environmental and health impacts on a global scale.

References

1. Covaci A, Gerecke AC, Law RJ, Voorspoels S, Kohler M, Heeb NV, Leslie H, Allchin CR, de Boer J (2006) Hexabromocyclododecanes (HBCDs) in the environment and humans: a review. Environ Sci Technol 40(12):3679–3688
2. Law RJ, Covaci A, Harrad S, Herzke D, Abdallah MA-E, Fernie K, Toms L-ML, Takigami H (2014) Levels and trends of PBDEs and HBCDs in the global environment: status at the end of 2012. Environ Int 65:147–158
3. Cao X, Lu Y, Zhang Y, Khan K, Wang C, Baninla Y (2018) An overview of hexabromocyclododecane (HBCDs) in environmental media with focus on their potential risk and management in China. Environ Pollut 236:283–295
4. Marvin CH, Tomy GT, Armitage JM, Arnot JA, McCarty L, Covaci A, Palace V (2011) Hexabromocyclododecane: current understanding of chemistry, environmental fate and toxicology and implications for global management. Environ Sci Technol 45(20):8613–8623
5. Heeb NV, Schweizer WB, Kohler M, Gerecke AC (2005) Structure elucidation of hexabromocyclododecanes—a class of compounds with a complex stereochemistry. Chemosphere 61(1):65–73
6. Zhang Y, Li Q, Lu Y, Jones K, Sweetman AJ (2016) Hexabromocyclododecanes (HBCDDs) in surface soils from coastal cities in North China: correlation between diastereoisomer profiles and industrial activities. Chemosphere 148:504–510
7. Zhu H, Zhang K, Sun H, Wang F, Yao Y (2017) Spatial and temporal distributions of hexabromocyclododecanes in the vicinity of an expanded polystyrene material manufacturing plant in Tianjin, China. Environ Pollut 222:338–347
8. Manø S, Herzke D, Schlabach M (2008) New organic pollutants in Air, 2007 (Norwegian Pollution Monitoring Programme, TA 2689/2010). Climate and Control Agency of Norway, Oslo
9. Olukunle OI, Venier M, Hites RA, Salamova A (2018) Atmospheric concentrations of hexabromocyclododecane (HBCDD) diastereomers in the Great Lakes region. Chemosphere 200:464–470
10. Yu Z, Chen L, Mai B, Wu M, Sheng G, Fu J, Peng P (2008) Diastereoisomer-and enantiomer-specific profiles of hexabromocyclododecane in the atmosphere of an urban city in South China. Environ Sci Technol 42(11):3996–4001
11. Wu M, Han T, Xu G, Zang C, Li Y, Sun R, Xu B, Sun Y, Chen F, Tang L (2016) Occurrence of hexabromocyclododecane in soil and road dust from mixed-land-use areas of Shanghai, China, and its implications for human exposure. Sci Total Environ 559:282–290
12. Su G, McGoldrick DJ, Clark MG, Evans MS, Gledhill M, Garron C, Armelin A, Backus SM, Letcher RJ (2018) Isomer-specific hexabromocyclododecane (HBCDD) levels in top predator fish from across Canada and 36-year temporal trends in Lake Ontario. Environ Sci Technol 52(11):6197–6207
13. Arnot JA, McCarty L, Armitage JM, Toose-Reid L, Wania F, Cousins IT (2009) An evaluation of hexabromocyclododecane (HBCD) for persistent organic pollutant (POP) properties and the potential for adverse effects in the environment. Submitted to European Brominated Flame Retardant Industry Panel (EBFRIP). University of Toronto Scarborough, Toronto, Canada
14. Heeb NV, Bernd Schweizer W, Mattrel P, Haag R, Gerecke AC, Schmid P, Zennegg M, Vonmont H (2008) Regio- and stereoselective isomerization of hexabromocyclododecanes (HBCDs): Kinetics and mechanism of γ- to α-HBCD isomerization. Chemosphere 73(8):1201–1210
15. Harrad S, Abdallah MA-E, Covaci A (2009) Causes of variability in concentrations and diastereomer patterns of hexabromocyclododecanes in indoor dust. Environ Int 35(3):573–579
16. Peled M, Scharia R, Sondack D (1995) Thermal rearrangement of hexabromocyclododecane (HBCD). In: Desmurs J-R, Gérard B, Goldstein MJ (eds) Industrial chemistry library, advances in organobromine chemistry. Elsevier, Amsterdam, The Netherlands, pp 92–99
17. Zhao Y, Zhang X, Sojinu OS (2010) Thermodynamics and photochemical properties of α, β, and γ-hexabromocyclododecanes: a theoretical study. Chemosphere 80(2):150–156

18. Heeb NV, Graf H, Bernd Schweizer W, Lienemann P (2010) Thermally-induced transformation of hexabromocyclo dodecanes and isobutoxypenta bromocyclododecanes in flame-proofed polystyrene materials. Chemosphere 80(7):701–708
19. Köppen R, Becker R, Jung C, Nehls I (2008) On the thermally induced isomerisation of hexabromocyclododecane stereoisomers. Chemosphere 71(4):656–662
20. Ni H, Lu S, Mo T, Zeng H (2016) Brominated flame retardant emissions from the open burning of five plastic wastes and implications for environmental exposure in China. Environ Pollut 214:70–76
21. MacLeod M, Scheringer M, Hungerbühler K (2007) Estimating enthalpy of vaporization from vapor pressure using Trouton's rule. Environ Sci Technol 41(8):2827–2832
22. Li L, Wania F (2018) Elucidating the variability in the hexabromocyclododecane diastereomer profile in the global environment. Environ Sci Technol 52(18):10532–10542
23. Cai W, Wan L, Jiang Y, Wang C, Lin L (2015) The short-lived buildings in China: impacts on water, energy and carbon emissions. Environ Sci Technol 49(24):13921–13928
24. Verbiest P, van den Ven P (1997) Measurement of capital stock and consumption of fixed capital in the Netherlands. STD/NA(97)12. Statistics Netherlands, The Hague, The Netherlands
25. Bergsdal H, Bohne RA, Brattebø H (2007) Projection of construction and demolition waste in Norway. J Ind Ecol 11(3):27–39
26. Abbasi G, Buser AM, Soehl A, Murray MW, Diamond ML (2015) Stocks and flows of PBDEs in products from use to waste in the US and Canada from 1970 to 2020. Environ Sci Technol 49(3):1521–1528
27. Tsiliyannis CA (2005) Dynamic modelling of packaging material flow systems. Waste Manag Res 23(2):155–166
28. OECD (2008) Measurement of Depreciation Rates Based on Disposal Asset Data in Japan. Working party on national accounts. Organisation for Economic Co-operation and Development (OECD), Paris
29. United Nations Commodity Trade Statistics Database (Comtrade) (2017)
30. Statistics of US Business (SUSB) (2008–2012 Annual Datasets) (2017)
31. United Nations Industrial Commodity Statistics Database (2017) United Nations Statistics Division
32. China National Chemical Information Center (1997–2016) China Chemical Industry Yearbook 1997–2016. China Petroleum & Chemical Industry Federation, Beijing
33. Breivik K, Armitage JM, Wania F, Jones KC (2014) Tracking the global generation and exports of e-waste. Do existing estimates add up? Environ Sci Technol 48(15):8735–8743
34. UNEP (2017) Revised Draft Guidance for the Inventory of Hexabromocyclododecane (HBCD). United Nations Environment Programme, Nairobi
35. VECAP (2013) The voluntary emissions control action program: European annual progress report 2013. voluntary emissions control action program (VECAP) in Europe, Brussels, Belgium
36. EU (2008) Risk assessment on hexabromocyclododecane (Final report). European Union (EU), Luxembourg
37. Kajiwara N, Desborough J, Harrad S, Takigami H (2013) Photolysis of brominated flame retardants in textiles exposed to natural sunlight. Environ Sci Process Impacts 15(3):653–660
38. ESWI (2011) Study on waste related issues of newly listed POPs and candidate POPs (Final Report). Consortium Expert team to Support Waste Implementation (ESWI), Munich
39. Abdallah MA-E, Sharkey M, Berresheim H, Harrad S (2018) Hexabromocyclododecane in polystyrene packaging: a downside of recycling? Chemosphere 199:612–616
40. Rani M, Shim WJ, Han GM, Jang M, Song YK, Hong SH (2014) Hexabromocyclododecane in polystyrene based consumer products: an evidence of unregulated use. Chemosphere 110:111–119
41. Gao S, Wang J, Yu Z, Guo Q, Sheng G, Fu J (2011) Hexabromocyclododecanes in surface soils from e-waste recycling areas and industrial areas in South China: concentrations, diastereoisomer-and enantiomer-specific profiles, and inventory. Environ Sci Technol 45(6):2093–2099

42. Breivik K, Armitage JM, Wania F, Sweetman AJ, Jones KC (2016) Tracking the global distribution of persistent organic pollutants accounting for e-waste exports to developing regions. Environ Sci Technol 50(2):798–805
43. Kajiwara N, Sueoka M, Ohiwa T, Takigami H (2009) Determination of flame-retardant hexabromocyclododecane diastereomers in textiles. Chemosphere 74(11):1485–1489
44. MacLeod M, von Waldow H, Tay P, Armitage JM, Wöhrnschimmel H, Riley WJ, McKone TE, Hungerbuhler K (2011) BETR global—a geographically-explicit global-scale multimedia contaminant fate model. Environ Pollut 159(5):1442–1445
45. Le TT, Son M-H, Nam I-H, Yoon H, Kang Y-G, Chang Y-S (2017) Transformation of hexabromocyclododecane in contaminated soil in association with microbial diversity. J Hazard Mat 325:82–89
46. Davis JW, Gonsior SJ, Markham DA, Friederich U, Hunziker RW, Ariano JM (2006) Biodegradation and product identification of 14C hexabromocyclododecane in wastewater sludge and freshwater aquatic sediment. Environ Sci Technol 40(17):5395–5401
47. Gerecke AC, Giger W, Hartmann PC, Heeb NV, Kohler H-PE, Schmid P, Zennegg M, Kohler M (2006) Anaerobic degradation of brominated flame retardants in sewage sludge. Chemosphere 64(2):311–317
48. Lee SC, Sverko E, Harner T, Pozo K, Barresi E, Schachtschneider J, Zaruk D, DeJong M, Narayan J (2016) Retrospective analysis of "new" flame retardants in the global atmosphere under the GAPS network. Environ Pollut 217:62–69
49. Rauert C, Schuster JK, Eng A, Harner T (2018) Global atmospheric concentrations of brominated and chlorinated flame retardants and organophosphate esters. Environ Sci Technol 52(5):2777–2789
50. Chen D, Hale RC, La Guardia MJ, Luellen D, Kim S, Geisz HN (2015) Hexabromocyclododecane flame retardant in Antarctica: research stations as sources. Environ Pollut 206:611–618
51. Newton S, Sellström U, de Wit CA (2015) Emerging flame retardants, PBDEs, and HBCDDs in indoor and outdoor media in Stockholm, Sweden. Environ Sci Technol 49(5):2912–2920
52. Roosens L, Abdallah MA-E, Harrad S, Neels H, Covaci A (2009) Exposure to hexabromocyclododecanes (HBCDs) via dust ingestion, but not diet, correlates with concentrations in human serum: preliminary results. Environ Health Perspect 117(11):1707–1712
53. Zegers BN, Mets A, van Bommel R, Minkenberg C, Hamers T, Kamstra JH, Pierce GJ, Boon JP (2005) Levels of hexabromocyclododecane in harbor porpoises and common dolphins from Western European Seas, with evidence for stereoisomer-specific biotransformation by cytochrome P450. Environ Sci Technol 39(7):2095–2100
54. Du M, Lin L, Yan C, Zhang X (2012) Diastereoisomer- and enantiomer-specific accumulation, depuration, and bioisomerization of hexabromocyclododecanes in Zebrafish (*Danio rerio*). Environ Sci Technol 46(20):11040–11046
55. Zhang Y, Sun H, Ruan Y (2014) Enantiomer-specific accumulation, depuration, metabolization and isomerization of hexabromocyclododecane (HBCD) diastereomers in mirror carp from water. J Hazard Mat 264:8–15

Chapter 6
Effective Management of Demolition Waste Containing Hexabromocyclododecane in China

Abstract Demolition waste containing hexabromocyclododecane (HBCDD) can be disposed of at either a waste, material, or substance level; however, their efficiency in mitigating long-term HBCDD contamination can be different. In this chapter, we quantify how much HBCDD emissions can be avoided and how efficiently the stock of HBCDD embedded in the demolition waste can be reduced if mainland China adopts different levels of waste management. We find that a pre-demolition screening combined with environmentally sound treatment, that is, management at the substance level, is the most effective end-of-life management option for minimizing HBCDD emissions and stocks. This level of waste management reduces slightly more HBCDD emissions than accelerating the ban of backfill or illegal open dumping of general demolition waste, that is, management at the waste level. While increasing the recycling of polystyrene materials, that is, management of waste at the material level, is ideal for the circular economy, it is least effective in reducing HBCDD emissions and may introduce this problematic chemical into recovered materials as new in-use stocks.

6.1 Introduction

Chapter 5 shows that, in addition to immediate and concentrated emissions from production and new uses, there are also long-term, continuous and dispersive emissions of HBCDD from its in-use and end-of-life waste (mainly demolition waste) stocks, if appropriate emission abatement measures are absent in the coming decades. In general, we can minimize emissions of a toxic chemical from its waste stock through three levels of end-of-life management strategies:

(i) Appropriate management of all municipal or hazardous wastes that possibly contain the chemical without identification or separation of chemical-containing waste (i.e., at a *waste* level);

(ii) Separation, followed by recycling, irreversible destruction, or disposal of the materials that possibly contain the chemical without distinguishing between chemical-containing and chemical-free materials (i.e., at a *material* level); and

© Springer Nature Singapore Pte Ltd. 2020
L. Li, *Modeling the Fate of Chemicals in Products*, Springer Theses,
https://doi.org/10.1007/978-981-15-0579-9_6

(iii) Identification of the chemical-containing materials, followed by separation and irreversible destruction (i.e., at a *substance* level).

Examples of the first strategy include to ban backfilling or illegal open dumping of municipal solid waste, and incineration of municipal or hazardous solid waste for thermal recovery. This strategy does not discriminate toxic chemicals from general solid waste and requires no specific information on materials (products) or chemicals. In this sense, such a "one-size-fits-all" strategy is crude and has nothing to do with the context of "chemicals in products (CiPs)". The second practice meets the requirement of the "circular economy" because the recovery and reuse of used materials minimize the inputs of raw materials and energies into the manufacture of new materials. This practice includes, for instance, re-bonding, re-grinding and molding of polymeric materials, and then reusing the recycled materials in new production. It requires information on materials (products) in which the use of toxic chemicals is found. However, this practice would be problematic if a hazardous chemical were incompletely destroyed and hence remained in recycled materials. The third strategy builds on discrimination of chemical-containing products from their chemical-free counterparts, which often requires chemical-specific knowledge, as well as screening or separation techniques. For instance, X-ray fluorescence spectroscopy (XRF) is used as a rapid and simple screening technique for detecting the presence of HBCDD-bound bromine for old products [1, 2]. The third strategy is the most delicate, but meanwhile the most expensive, among the three.

All three end-of-life management strategies are potentially viable for controlling and abating HBCDD emissions from waste stocks in a region, e.g., in mainland China. Before regional policies are made and implemented, we need to quantify how much HBCDD emissions can be avoided and how efficiently the stock of HBCDD embedded in the demolition waste can be reduced. If additional information on the economic cost is provided, we can find out the most cost-effective end-of-life management strategy to tackle the demolition waste. To this end, we will compare how three candidate end-of-life management strategies perform in terms of migrating the long-term HBCDD emissions in mainland China, using a scenario-based substance flow analysis. A single region, instead of multiple regions, is selected for modeling because in so doing can we exclude the influence from variations in other socioeconomic conditions such as the transboundary trade of waste and the waste disposal status, which, therefore, renders the comparison more transparent and explicit.

6.2 Methods and Data

6.2.1 Modifications for the China-Specific Situation

In this chapter, the CiP-CAFE model is slightly modified to facilitate the simulation of the anthropospheric fate of technical HBCDD mixture in mainland China (Region 1). In mainland China, the technical HBCDD mixture is used in EPS-insulation boards (AP1) and XPS-insulation boards (AP2) in buildings, and textile coating agent (AP3). A national market survey indicates that HBCDD has not been used in EPS-package (AP4) and high impact polystyrene (HIPS) in electrical and electronic equipment (AP5) due to its price disadvantage over other flame retardants [3].

Most parameters, including annual production, interregional trades, emission and waste factors, and waste disposal ratios, remain the same as those used for Region 1 in Chap. 5, except for the following:

First, instead of targeting at three HBCDD diastereomers in Chap. 5, here we focus on the technical HBCDD mixture (CAS no. 25637-99-4 or 3194-55-6). For the technical HBCDD mixture, thermodynamically consistent partition coefficients at 25 °C are taken from Arnot et al. [4]. Internal energies of phase transfer, ΔU_W, ΔU_A, and ΔU_O, which are used to adjust the partition coefficients for temperature dependence, are calculated in accordance with MacLeod et al. [5] based on Trouton's Rule and defaults. Degradation half-lives in soil (and solid waste in landfill, and dumping and simple landfill), water (and wastewater) and air at 25 °C are assumed to be the same as those of HBCDD diastereomers (see Chap. 5). Activation energies for adjusting the degradation half-lives for temperature dependence are defaults in the CiP-CAFE model.

Second, the original CiP-CAFE model assumes that outflows from "recycling and reuse (WD5)" stage, except for the fraction decomposed, enter the "industrial processes (LC2)" and are then distributed among all applications. Here we adopt a more sophisticated assumption that 20% of the outflows are reused in XPS-insulation boards (AP2) (Q. Meng, personal communication), 10% to produce calcic-plastic composite materials (i.e., outside the modeled system) (Z. Zhou, personal communication), and the remaining 70% are not recyclable by current technologies and hence subject to landfill (WD1) [6]. This assumption is believed to reflect better the realistic situation in mainland China.

Third, we assume that all, rather than a fraction of, generated wastewater is directly discharged into the hydrosphere because the majority (>80%) of the Chinese producers are small or microenterprises (i.e., primitive workshops) without appropriate wastewater treatment facilities.[1] In addition, we assume that the emissions from disposal of industrial waste (e.g., from packaging waste residues) are negligible because, in mainland China, industrial waste is often either incinerated, during which HBCDD is destroyed, or collected for reuse [7].

6.2.2 Disposal Scenarios of HBCDD-Containing Demolition Waste

We focus on HBCDD-containing demolition waste arising from EPS- (AP1) and XPS-insulation boards (AP2) at the end of building lifespan.

---

[1] Surveyed data presented in the project "Developing Emission Inventory and Designing Implementation Strategy for Regulating HBCDD in China" conducted by Beijing Normal University.

To begin with, we review the status quo and possible future improvement of the disposal of HBCDD-containing demolition waste in China. At the *substance* level, HBCDD-specific disposal has never been mandated in China, because cheap and rapid screening techniques (e.g., X-ray fluorescence equipment) are currently not accessible and affordable to common Chinese users. It is unlikely that China will adopt HBCDD-specific disposal in the coming years in a business-as-usual situation. At the *material* level, recycling of polystyrene materials is rather rare: 1.25% and 5% of the total demolition waste was subject to recycling in 2011 and 2013, respectively [8]. We can simply assume that the waste disposal ratio for WD5 increased by two percentage points each year from 2011 to 2013. A recent government document [9] announced that China will make greater efforts than ever before on the "comprehensive utilization" of demolition waste in the coming years; as such, the recycling rate of demolition waste is anticipated to increase at the same rate as that between 2011 and 2013. However, the increase in the recycling rate can hardly be infinite due to technological feasibilities and cost constraints; the recycling rate will likely plateau after reaching a ceiling of 30%, given the case in other countries, e.g., in Japan [10]. Finally, at the *waste* level, ~15% of the total generated demolition waste nationwide is landfilled (i.e., WD1—landfill) annually and the rest ~80% is subject to backfill or illegal open dumping (i.e., WD3—dumping and simple landfill). China is making every endeavor to eliminate backfilling or illegal open dumping of demolition waste; the fraction of demolition waste subject to backfill or illegal open dumping is declining by one percentage point each year (see a summary of currently available statistical data in Li et al. [11]).

Based on the information above, we define a business-as-usual "baseline" scenario (**Scenario I**; Fig. 6.1; Table 6.1), in which China is assumed to continue its current end-of-life management strategies on demolition waste.

Because the objective of this chapter is to compare the performance of end-of-life management strategies of different level, we additionally define three "alternative" scenarios for the future (Fig. 6.1; Table 6.1):

(i)   **Scenario II**, targeted at addressing the HBCDD issue at a waste level, assumes that China will accelerate the ban of backfilling or illegal open dumping of general demolition waste, and redirect the demolition waste being currently backfilled or dumped to engineered landfills. In other words, the waste distribution ratio for WD3 is shrinking at a rate faster than that between 2011 and 2013, and the waste distribution ratio for WD1 is increasing at the same pace as the decrease in WD3;

(ii) **Scenario III**, representing a material level, assumes that China will separate polystyrene materials from other demolition waste using appropriate techniques, and increase the recycling rate of polystyrene materials (i.e., the waste distribution ratio for WD5). Since polystyrene materials are usually recycled at a temperature [12] far below the decomposition temperature of HBCDD (240–270 °C [13]), HBCDD remains in the recycled polystyrene materials and is reused in new XPS-insulation boards;

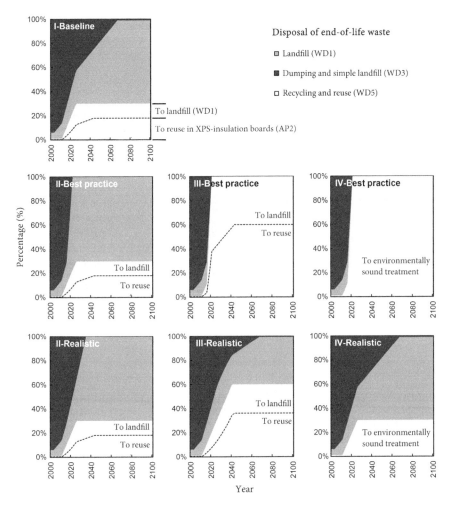

Fig. 6.1 Diagrammatic definition of the shares of waste management strategies in the baseline scenario (I) and three alternative scenarios (II to IV) Adapted with permission (Li et al. [11]) Copyright (2016) Elsevier

(iii) **Scenario IV**, aiming to mitigate HBCDD emissions from a substance per-spective, assumes that China will screen, separate and destroy the HBCDD-containing insulation boards using appropriate environmentally sound tech-niques. HBCDD is irreversibly destroyed during environmentally sound treat-ment (i.e., enters WD7).

Each alternative scenario comprises two sub-scenarios. The *best practice sub-scenario* includes ambitious, rapid but rather unrealistic mitigation of HBCDD emis-sions and assumes that all these actions can be accomplished within China's forth-coming five-year implementation period of the Stockholm Convention (2017–2021).

Table 6.1 Summarized definitions of baseline scenario (I) and three alternative scenarios (II to IV) of end-of-life management for the simulation period 2016–2100

Scenario	Waste management measures	Sub-scenario	
		(a) Best practice	(b) Realistic
I. Baseline	No additional measures (business-as-usual)		
II. Accelerating the ban of backfilling or illegal open dumping of demolition waste	To bring forward the complete prohibition of backfilling or illegal open dumping of general demolition waste	Linear increase in the share of the controlled landfill to substitute for backfilling or illegal open dumping until a complete substitution in 2021	Linear increase in the share of the controlled landfill to substitute for backfilling or illegal open dumping until a complete substitution in 2035
III. Increasing recycling of polystyrene materials	To improve the recycling rate of polystyrene materials, but not to increase the fraction of polystyrene materials reused in new XPS-insulation boards	Linear increase in the annual recycling rate of polystyrene materials to 100% from 2021 onwards	Linear increase in the annual recycling rate of polystyrene materials to a doubled ceiling of 60% from 2041 onwards
IV. Screening and incinerating HBCDD-containing insulation boards	To scan, separate and incinerate the HBCDD-containing insulation boards	Pre-demolition screening of HBCDD-containing insulation boards	Post-recovery screening of HBCDD-containing insulation boards

By contrast, the *realistic sub-scenario* represents a possible route towards the environmentally sound management of HBCDD-containing demolition waste, in line with the current trend in China's national implementation plan to the Stockholm Convention. For each scenario, the effectiveness of the emission mitigation is deemed to be robust if a consistent mitigation trend appears in the two sub-scenarios with different specific input values assumed.

6.2.3 Evaluation of Emission Mitigation Performance

Two indicators are calculated for evaluating the emission mitigation of alternative scenarios:

(i) The *mitigation potential* means the cumulative HBCDD emissions avoided by an end-of-life management strategy if it is enforced at a realistic pace. It is defined as the difference in cumulative emissions between the *baseline* scenario and the *realistic* sub-scenario from 2016 to 2100;

(ii) The *improvability* means the additional amount of avoided HBCDD emissions if the management becomes more stringent relative to that in each realistic sub-scenario. It is defined as the difference in cumulative emissions between *best practice* and *realistic* sub-scenarios from 2016 to 2100.

For facilitating comparison, we normalize the emission mitigation potential and improvability by the cumulative emissions in the baseline scenario from 2016 to 2100.

6.3 Overview of Stocks and Emissions in Mainland China

Prior to exploring the performance of different end-of-life management strategies in reducing the long-term HBCDD emissions, we first have an overview of the estimated in-use and waste stocks, and emissions of HBCDD in mainland China (Region 1), in the baseline scenario assuming that no emission mitigation measures are taken.

Figure 6.2 presents an overview of the estimated stocks. As shown in Fig. 5.1b in Chap. 5, massive production and uses of HBCDD in Region 1 began around 2000, flourished after 2005, and is projected to cease in 2021. By 2021, the estimated cumulative production will amount to 238 (216–260) kilotonnes (kt). Approximately 75% of the produced HBCDD is consumed domestically, resulting in an ever-increasing in-use stock (Fig. 6.2). The total in-use stock is estimated to peak at 178 (137–192) kt in 2021 when the intended use of HBCDD in virgin insulation boards is scheduled to stop. By then, 73 and 26% of the total in-use HBCDD will reside in the installed

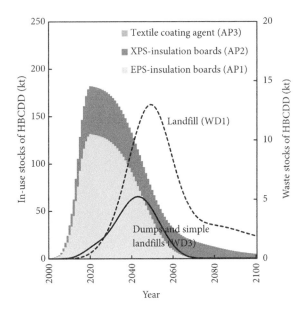

Fig. 6.2 Temporal evolution of in-use and waste stocks of HBCDD in China in the baseline scenario that assumes no additional waste management measures in the future. Reproduced with permission (Li et al. [11]) Copyright (2016) Elsevier

EPS (AP1) and XPS-insulation boards (AP2), respectively, making the in-use stock of EPS-insulation boards the largest reservoir of HBCDD; in comparison, HBCDD storage in fabric and textile products (AP3) will be minor. The depletion of the in-use stock is going to accelerate around 2030 when the demolition of buildings with HBCDD-containing insulation boards increases nationwide. The depletion of the in-use XPS stock is slightly slower than that of in-use EPS stock (Fig. 6.2) because a fraction of HBCDD will return to new XPS-insulation boards by the inadvertent recycling of unidentifiable HBCDD-containing waste. This substance circulation makes the in-use XPS stock the largest reservoir after the 2050s. The misalignment of curves between the in-use and two waste stocks demonstrates that the response of waste to new use is temporally buffered (Fig. 6.2). A higher proportion of HBCDD would flow to dumping and simple landfill (WD3) before 2025, but landfill (WD1) will dominate the waste stock afterward (Fig. 6.2), because it will probably take some time for the Chinese administrators to take legislative and technical measures to control the current widespread illegal demolition waste treatment methods.

Figure 6.3 displays the estimated annual emissions of HBCDD from individual sources. The estimated annual emissions of HBCDD comprise two peaks. The first peak mainly results from production and industrial processes with a maximum of ~8 t year^{-1} around 2015; while the second peak represents the disposal of end-of-life waste with a maximum of ~3 t year^{-1} around 2045. The current significance of industrial HBCDD emission sources in China is evident from numerous extreme concentrations in the vicinity of manufacturing and processing sites of HBCDD and HBCDD-containing products [14, 15].

The end-of-life waste is a big issue in the future. For the entire simulation period 2000–2100, the disposal of end-of-life waste contributes the largest share (49%) to

Fig. 6.3 Annual emissions of HBCDD in China from production, industrial processes, in service, and disposal of end-of-life waste in the baseline scenario that assumes no additional waste management measures in the future. Reproduced with permission (Li et al. [11]) Copyright (2016) Elsevier

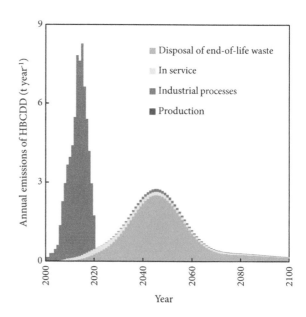

cumulative HBCDD emissions, which is even higher than industrial processes (44%). Demolition waste overwhelmingly dominates the end-of-life waste. Our calculation indicates that these waste stocks are not the final, permanent destinations for HBCDD; there will be lasting, robust and stable HBCDD streams "bleeding" from the waste stocks to the environment over many more decades. In particular, it is only during the past two decades that we have witnessed the intensive uses of HBCDD in China yet a whole century will be needed to eradicate HBCDD from its waste stocks (Figs. 6.2 and 6.3). Therefore, appropriate management of demolition waste is essential for minimizing the future HBCDD emissions.

6.4 Emission Mitigations in Different End-of-Life Disposal Scenarios

Using a scenario analysis, we project HBCDD emissions (Fig. 6.4a), in-use stocks (Fig. 6.4b) and the waste stocks (Fig. 6.4c) from 2016 to 2100, to evaluate the performance of each possible end-of-life management strategy in mitigating HBCDD contamination.

An overview of the simulation results reveals that all three proposed end-of-life management strategies are capable of abating future HBCDD emissions (Table 6.2). However, HBCDD emissions in China will most likely keep growing (before 2040), as an ascending trend in the annual emissions is common in all six alternative sub-scenarios (Fig. 6.4a).

Compared with the baseline scenario, accelerating the ban of backfill or illegal open dumping of general demolition waste will lower the cumulative emissions of

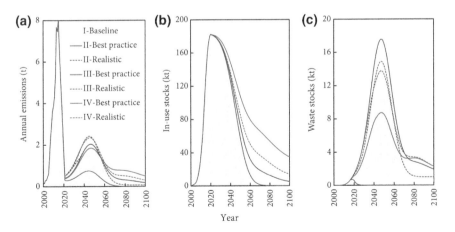

Fig. 6.4 Temporal evolution of annual emissions (**a**), in-use stocks (**b**) and waste stocks (**c**) of HBCDD in China in the baseline (gray shades) and three alternative scenarios (colorful curves). Reproduced with permission (Li et al. [11]) Copyright (2016) Elsevier

Table 6.2 Cumulative avoided emissions between 2016 and 2100 in the alternative scenarios, expressed as percentages in the cumulative emission between 2016 and 2100 in the baseline scenario

	Scenario-II (At the waste level)	Scenario-III (At the material level)	Scenario-IV (At the substance level)
Mitigation potential	15%	2%	16%
Improvability	3%	5%	45%

HBCDD by 18% in Scenario II-a and 15% in Scenario II-b (Fig. 6.4). Therefore, the mitigation potential is 15% and the improvability is 3% (Table 6.2). However, the in-use and waste stocks remain almost unchanged, which indicates that such an end-of-pipe improvement does not affect the behavior of HBCDD in the lifecycle, but merely redirects the waste streams from a high-emission backfill or illegal open dumping stock to a low-emission controlled landfill stock. However, the both stocks are intermediate, rather than permanent, sinks in the anthroposphere and still release HBCDD into the environment. The lowered emissions imply that a higher fraction of HBCDD is merely (and transiently) stored in landfill sites without relevance to the immobilization or elimination. With appropriate aftercare, the controlled landfill can be a clean disposal measure; however, it leaves a large number of contaminated sites, in which liners and leachate collection systems will degrade over time and ultimately fail to contain contaminants, and thus further releases will occur, or further expensive treatment and remediation will be needed [16, 17]. Furthermore, given that the improvability of emission mitigation is small (Table 6.2), we cannot expect a substantial increment in emission mitigation by simply implementing a more stringent regulation on general demolition waste than business as usual.

Increasing the recycling of polystyrene materials leads to the smallest reduction in the cumulative HBCDD emissions among the three end-of-life management strategies (Table 6.2 and Fig. 6.4a). The mitigation potential is 2% and the improvability is 5% (Table 6.2). In other words, this end-of-life management strategy is not efficient in reducing HBCDD emissions, although it is believed to be beneficial for the circular economy in the perspective of maximizing the use of materials. Without visual identification system for removing HBCDD-containing insulation boards from total polystyrene waste, the indiscriminate recovery reinstates HBCDD into the XPS product chain and extends the residence time of HBCDD in in-use stocks (Fig. 6.4b), thus delaying the flow of HBCDD streams into waste stocks (Fig. 6.4c). Compared with other end-of-life management strategies, this strategy leads to higher emissions in the best practice (III-a) than the realistic (III-b) sub-scenario from 2057 onwards. This is due to that HBCDD is degraded much slower in end-products (e.g., no observable degradation in textiles in a 371-day experiment [18]) than in landfill or backfill (e.g., an aerobic degradation half-life of 63 days [19]). HBCDD would be released to some extent from recycled products in the same way as in virgin products [20], thus prolonging consumer exposure to this compound. Moreover, consumer exposure would be of concern if the recovered HBCDD-containing polystyrene goes into packaging and food contact materials [21, 22].

Post-recovery screening and environmentally sound treatment of HBCDD-containing insulation boards (Scenario IV-b) reduce 16% of cumulative HBCDD emissions, which is comparable to the reduction in Scenario II-b (Table 6.2). Combining pre-demolition screening and environmentally sound treatment (Scenario IV-a) additionally reduces another 45% of cumulative HBCDD emissions (Table 6.2), which makes it the most effective measure in achieving emission reduction goals. In addition, up to 98% of the in-use stock and all waste stock is reduced in Scenario IV-a (Fig. 6.4c). Therefore, the earlier identification and destruction of the HBCDD constituent, the less adverse influences the anthroposphere and the environment will suffer. However, despite their highest effectiveness in controlling the HBCDD contamination, both screening (e.g., X-ray fluorescence spectroscopy, or labeling before marketing) and irreversible destruction techniques have been in limited use in most countries. To make this technique viable, we still need further assessments to scrutinize (i) its economic accessibility and the affordability of general Chinese deconstruction contractors and demolition waste disposal enterprises; and (ii) its technical practicability and effectiveness in detecting HBCDD constituent under most circumstances. Furthermore, the future large-scale implementation of this technique will also need an appropriate regulatory framework.

While we have illustrated the performance of emission mitigation in each scenario, the realistic situation in future China might be a combination of the three since the scenarios are not "mutually exclusive". For instance, a simultaneous combination of the realistic sub-scenarios of Scenarios II, III and IV (data not shown here) shows more effective in emission mitigation than the individuals, which reduces 28.4% (corresponding to 44.4 tonnes) of total emissions compared with the baseline scenario for the simulation period 2016–2100. Given that the sum of the emission mitigation potentials in the three individual realistic sub-scenarios is 33% (corresponding to 36 t) (Table 6.2), this combination yields synergistic emission mitigation of 8.4 tonnes. Meanwhile, other end-of-life management strategies can also be involved in future HBCDD-containing waste treatment since the three scenarios are not "collectively exhaustive". For example, the introduction of solvent-based separation technique helps to distinguish dissolvable polystyrene matrix and non-dissolved HBCDD constituent [23], which facilitates incinerating HBCDD alone and recycling polystyrene from HBCDD-containing insulation boards. In a sense, this method benefits both polystyrene resource management and emission abatement.

6.5 Summary

In this chapter, we evaluate the performance of the disposal of HBCDD-containing demolition waste at the waste, material, and substance levels in mitigating future HBCDD emissions in China, based on a scenario-based dynamic substance flow analysis. We reveal imminent concerns about the management of HBCDD-containing demolition waste in China. We find that a pre-demolition screening combined with environmentally sound treatment, i.e., management at the substance level, is the most effective end-of-life management option for minimizing emissions. Management at the waste level performs slightly worse than that at the substance level. While management of waste at the material level is ideal for the circular economy, it is least effective in reducing HBCDD emissions and may introduce this problematic chemical into recovered materials.

References

1. Schlummer M, Vogelsang J, Fiedler D, Gruber L, Wolz G (2015) Rapid identification of polystyrene foam wastes containing hexabromocyclododecane or its alternative polymeric brominated flame retardant by X-ray fluorescence spectroscopy. Waste Manag Res 33(7):662–670
2. Secretariat of the Basel Convention (2015) Technical guidelines for the environmentally sound management of wastes consisting of, containing or contaminated with hexabromocyclododecane. Secretariat of the Basel Convention, Geneva
3. Beijing Institute of Technology (2011) Survey report on basic information of HBCDD in China. Beijing
4. Arnot JA, McCarty L, Armitage JM, Toose-Reid L, Wania F, Cousins IT (2009) An evaluation of hexabromocyclododecane (HBCD) for persistent organic pollutant (POP) properties and the potential for adverse effects in the environment. Submitted to European brominated flame retardant industry panel (EBFRIP). University of Toronto Scarborough, Toronto, Canada
5. MacLeod M, Scheringer M, Hungerbühler K (2007) Estimating enthalpy of vaporization from vapor pressure using Trouton's rule. Environ Sci Technol 41(8):2827–2832
6. Anonym (2013) China should increase the use of obsolete PS foams. Plast Sci Technol 41(2):70 (in Chinese)
7. Peking University (2012) Socio-economic impact analysis of implementing the Stockholm Convention to regulate HBCDD. College of Environmental Sciences and Engineering, Peking University, Beijing
8. China's National Development and Reform Commission (2013, 2014) Annual report of comprehensive utilization of resources (2012 and 2014). China's National Development and Reform Commission (NDRC), Beijing
9. China's National Development and Reform Commission (2011) The twelfth five-year plan of comprehensive utilization of large industrial solid wastes. China's National Development and Reform Commission (NDRC), Beijing
10. Managaki S, Hondo H, Yokoyama Y, Miyake Y, Kobayashi T, Miyake A, Masunaga S (2010) Comparative study between life-cycle HBCD and CO_2 emissions for the risk trade-off analysis. Organohalogen Compd. 72:1642–1646
11. Li L, Weber R, Liu J, Hu J (2016) Long-term emissions of hexabromocyclododecane as a chemical of concern in products in China. Environ Int 91:291–300
12. Maharana T, Negi YS, Mohanty B (2007) Review article: recycling of polystyrene. Polym Plast Technol Eng 46(7):729–736
13. Barontini F, Cozzani V, Petarca L (2001) Thermal stability and decomposition products of hexabromocyclododecane. Ind Eng Chem Res 40(15):3270–3280
14. Li H, Zhang Q, Wang P, Li Y, Lv J, Chen W, Geng D, Wang Y, Wang T, Jiang G (2012) Levels and distribution of hexabromocyclododecane (HBCD) in environmental samples near manufacturing facilities in Laizhou Bay area, East China. J Environ Monit 14(10):2591–2597
15. Zhang W (2013) Production and market of expandable polystyrene (EPS) at home and abroad. China Elastomerics 23(6):76–80 (in Chinese with English abstract)
16. Shaw S, Blum A, Weber R, Kannan K, Rich D, Lucas D, Koshland CP, Dobraca D, Hanson S, Birnbaum LS (2010) Halogenated flame retardants: do the fire safety benefits justify the risks? Rev Environ Health 25(4):261–306

17. Weber R, Watson A, Forter M, Oliaei F (2011) Persistent organic pollutants and landfills—a review of past experiences and future challenges. Waste Manag Res 29(1):107–121

18. Kajiwara N, Desborough J, Harrad S, Takigami H (2013) Photolysis of brominated flame retardants in textiles exposed to natural sunlight. Environ Sci Process Impacts 15(3):653–660

19. Davis J, Gonsior S, Marty G, Ariano J (2005) The transformation of hexabromocyclododecane in aerobic and anaerobic soils and aquatic sediments. Water Res 39(6):1075–1084

20. USEPA (2014) Flame retardant alternatives for hexabromocyclododecane (HBCD) (Final report, EPA Publication 740R14001). U.S. Environmental Protection Agency, Washington D.C

21. Abdallah MA-E, Sharkey M, Berresheim H, Harrad S (2018) Hexabromocyclododecane in polystyrene packaging: a downside of recycling? Chemosphere 199:612–616

22. Rani M, Shim WJ, Han GM, Jang M, Song YK, Hong SH (2014) Hexabromocyclododecane in polystyrene based consumer products: an evidence of unregulated use. Chemosphere 110:111–119

23. Schlummer M, Mäurer A, Leitner T, Spruzina W (2006) Report: recycling of flame-retarded plastics from waste electric and electronic equipment (WEEE). Waste Manag Res 24(6):573–583

Printed in the United States
By Bookmasters